OREGON WILDLAND FIREFIGHTING

A HISTORY

SEAN DAVIS

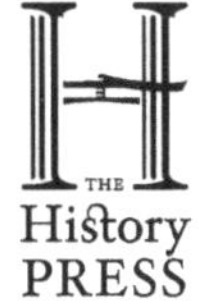

Published by The History Press
Charleston, SC
www.historypress.net

Front cover, top: Wildland firefighting instruction at Guard Training School in Tollgate, Oregon, in Umatilla National Forest in June 1939. *Photograph by Ray Filloon, courtesy of the U.S. Forest Service*; *bottom*: Just north of McDermitt, Oregon, a member of Fire Crew 7 keeps watch on the fire near the Blue Mountain Pass on the Long Draw Fire, 2012. This part of the fire was set by the crew in hopes of keeping the main part of the fire from interrupting the flow of traffic on Highway 95. *Photograph by Kevin Abel, Oregon BLM.*
Back cover, top: Umpqua NF Fires, 2017, Oregon. *Courtesy of the U.S. Forest Service–Pacific Northwest Region*; *bottom*: A group of wildland firefighters cool down a log, 1937. *Courtesy of the U.S. Forest Service.*

Opposite: The author and his son Cody Davis on the Flounce Fire in Southern Oregon, 2017. *Courtesy of Julio Molina.*

First published 2019

Manufactured in the United States

ISBN 9781467138505

Library of Congress Control Number: 2018960968

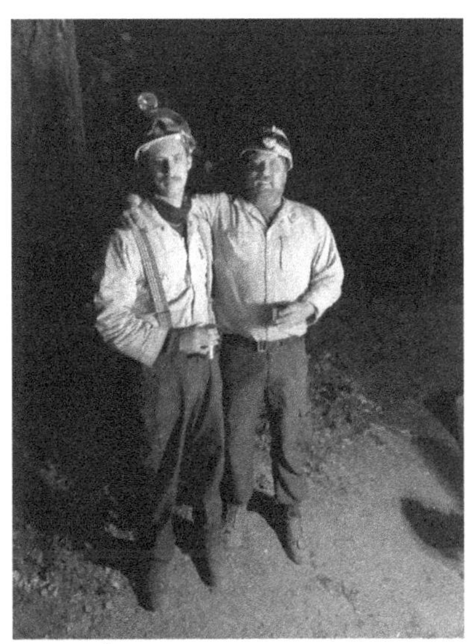

I'm dedicating this book to my son
Cody Davis. I couldn't be prouder of you.
Eric said you did a hell of a job out there.
I love you, son.

CONTENTS

ACKNOWLEDGEMENTS

When you're a wildland firefighter, you're part of a large family. For example, one day in January in 2018, I drove out to a small-town restaurant to interview Sam Swetland, a man who fought fires for over forty years for the U.S. Forest Service, and when I sat down across from him, he slid a photo of my aunt Jeanie to me. He and his wife, Cookie, knew my uncle and aunt through fighting fires. My uncle Ralph was happy and appreciative to see a new photograph of his late wife. Thanks for that, Sam, and thanks for a great interview. I also want to thank another fire family, Keith Hergert and Barbara Kuykendall. Without your support, I never would have written this book or come back to wildland firefighting. I'm appreciative of James Baker and our talk on our responsibilities to our environment. Thank you to Alan Maul, a historian at the Oregon Department of Forestry's Forest History Center. Thanks to Margaret Beilharz and Bill and Tyee Burwell. Thank you to Val Rapp and Eugene Skrine. Thanks to Damon Faust for talking me into this in the first place, and thanks to Eric Gatchell, Julio Molina and Dusty Webinger for teaching and working with me on the West Coast Wildland Engines. Fighting woodland fires is one of the hardest jobs I've ever done. I say that as someone who spent a career in the army infantry. But what is even more difficult is writing about wildland firefighting. I say that because so many people have lost their lives trying to protect other people, property and the land. I'm donating a part of the proceeds of this book to the Doug Dunbar Scholarship. Doug was a Prineville hotshot who lost his life near Storm King Mountain while fighting the South Canyon Fire near Glenwood Springs, Colorado. He died along with thirteen others. He started his career at what was then called the Blue River Ranger District. We went to the same school and played on the same baseball team here on the McKenzie River, where I now live. Working in the woods is a part of life here in the Cascades of Central Oregon. Like I say at the end of this book, it's in our DNA. Thank you to everyone who wears the greens and yellows.

INTRODUCTION

Anthropogenic. It's an impressive and semi-fancy word for human-caused wildfire. If you spent any time in a forest growing up, Smokey Bear will now likely appear in your mind waving his finger and saying, "Only you can prevent forest fires." Well, the truth is that man-made fires are both good and bad. Fire has always been a tool for mankind—in fact, it was one of its very first tools.

On one of his expeditions, David Douglas (the Scottish botanist who has a tree—the Douglas fir—and a Portland school district named after him) found that most of the tribes in the Upper Umpqua, Rogue River and Willamette Valley burned small patches of land every year. Douglas wrote, "Some of the natives tell me it is done for the purpose of urging the deer to frequent certain parts, to feed, which they leave unburned and of course they are easily killed. Others say that it is done in order that they might better find wild honey and grasshoppers, which both serve as articles of winter food."

In modern wildland firefighting, the tendency is to put out all fires instead of using some as preventive maintenance. Many say that this has disrupted the natural cycle, making our forests a massive natural disaster waiting to happen. The U.S. Forest Service website says as much: "Where fire was once a frequent visitor and served to keep forests open, growing and uncrowded, decades of fire suppression have created conditions far denser than they ought to be, making them more vulnerable to catastrophic fire. Nearly 2/3 of National Forest System lands have missed one or more expected fire cycles, resulting in elevated fire risk and a forest health concern for millions of acres."

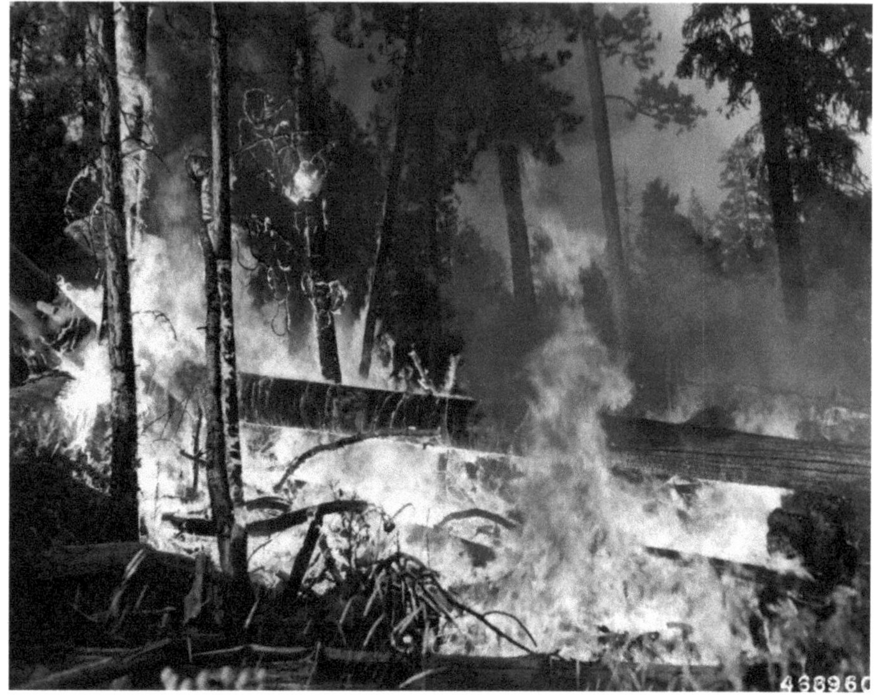

Forest fire on Wolf Creek in the Ochoco National Forest. Photograph taken by Bluford W. Muir, August 1951. *Courtesy of the U.S. Forest Service.*

Wildfires have raged across this land, devastating lives and property, since before Oregon became a state. In fact, the earliest written mention of a wildfire taking a person's life was noted by William Clark, of Lewis and Clark, during his expedition along the Oregon Trail. On October 29, 1804, Clark wrote,

> *The Prarie was Set on fire (or cought by accident) by a young man of the Mandins, the fire went with such velocity that it burnt to death a man & woman, who Could not get to any place of Safty, one man a woman & Child much burnt and Several narrowly escaped the flame. a boy half white was saved unhurt in the midst of the flaim. The couse of his being Saved was a Green buffalow Skin was thrown over him by his mother who perhaps had more fore Sight for the pertection of her Son, and* [l]*ess for herself than those who escaped the flame, the Fire did not burn under the Skin leaveing the grass round the boy. This fire passed our camp last* [night] *about 8 oClock P.M. it went with great rapitidity and looked Tremendious.*[1]

As pioneers and seekers of fortune poured into the region, they recorded some of the largest fires ever seen by man. The Yaquina Fire in 1849 started near what is now Corvallis, in the heart of the Willamette Valley, and raged until it hit the ocean at what is now Newport and Yaquina Bay. Imagine 480,000 acres of fire—to convert it into a measurement more easily visualized, it burned over fifty miles of fir, cedar, pine, birch, oak, alder and all the other flora and fauna sheltered in the area.

Then, in 1865, all hell broke loose, and the largest recorded fire in Oregon's history started just southeast of Salem; before it was over, it would burn over 1,540 square miles. That is almost a one-million-acre fire. That is five times the area of New York City burned to cinders. That is the Portland Metro area—times ten—on fire. In order for a fire to spread that far and be that hot, it needs to climb into the treetops.

When a fire starts to grow vertically as well as horizontally, wildland firefighters call this torching or crowning. When a single tree does this, it's called a candle.

The amount of violence and natural beauty on display when this occurs is hard to put into words. Imagine standing in the middle of a patch of earth the size of a football field scraped down to mineral soil. This safety zone is surrounded by the average Oregon forest, full of knee-high fescue and maybe some deer ferns or giant chain ferns. Below the big red cedars or Douglas firs are beds of discarded needles with red alder and Oregon iris flowers poking through.

A low thrumming cuts through the birdsong, and it grows. The volume keeps growing until you're convinced that there is some sort of locomotive on the other side of the tree line in front of you. What they say is true: it sounds like a freight train coming right at you, and now it's so loud that you believe it's right on top of you. Fear rises in the primal part of your brain because you can't believe anything natural can make a sound that loud.

A fire this big creates its own weather and is sucking needed oxygen into it in order to burn fuel while it simultaneously spreads, consumes and easily ignites everything around it. Sap inside trees boils and expands, causing some trees to pop, crack and even explode.

A dingy white smoke drains the sky of blue and darkens the morning. In fact, the sun is now only a pinhole in the sky above you. The air around you heats like a convection oven—exactly like a convection oven. It's so hot you can feel it down your throat, in your chest and into your stomach when you inhale. Flames crawl at you on the ground and jump from treetop to treetop above you. It's able to move so quickly because it is preheating the

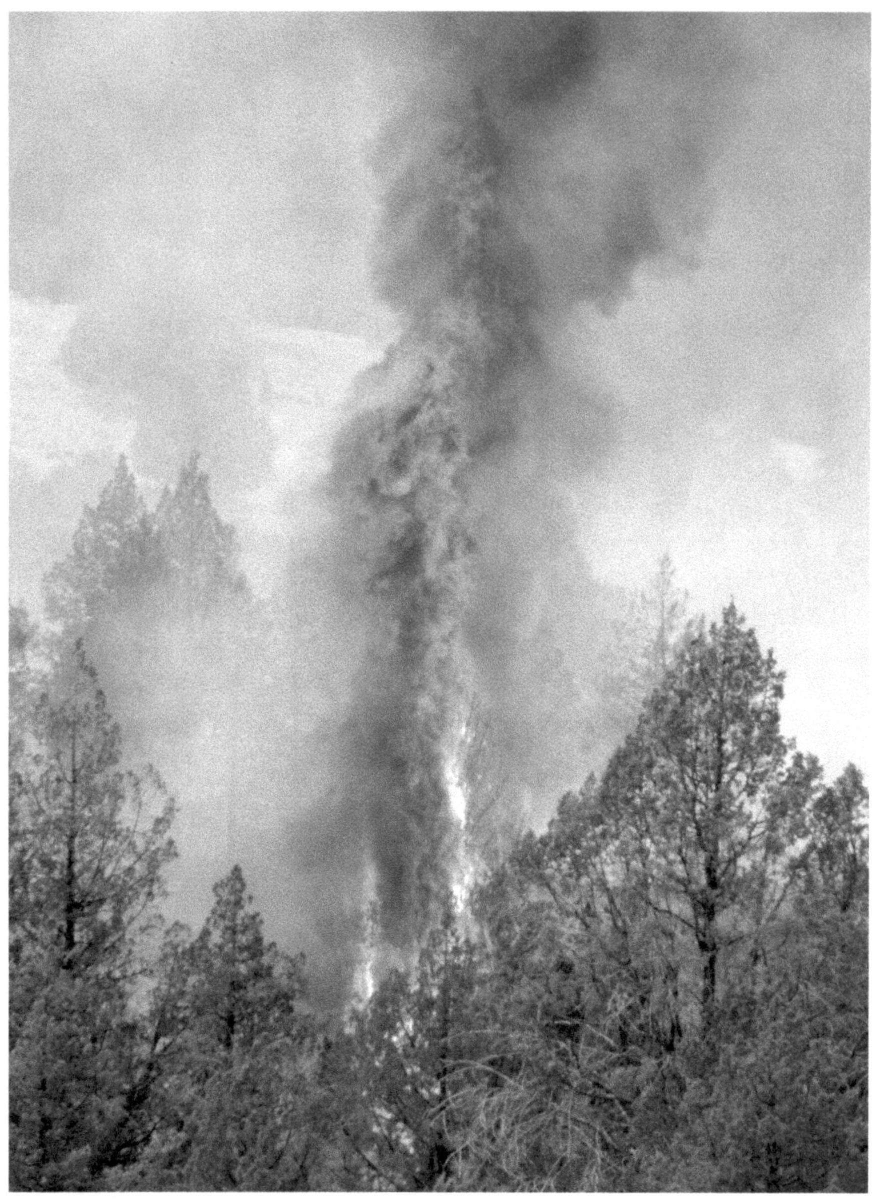

A fire moving vertically on the Government Flats Complex, 2013. *Courtesy of the U.S. Forest Service.*

fuel all around it, including you, completely drying everything out until it's all ready to ignite.

Then, there is a wall of flame in front of you, and it's fifty feet tall. No—not a wall of flame, because a wall is static. The fire thrashes, roars and consumes everything in its path. It surrounds you on both sides. Now, you struggle for breath. Greedy and unbelievably big, the fire steals the air from your lungs. You back away because the heat is unbearable, but now it circles your patch of dirt, and there's nowhere to go to make it stop.

According to the National Interagency Fire Center, only three out of the last eighty-six years did not include a wildland fire that claimed a human life. The NIFC chart, aptly titled "Wildland Fire Fatalities by Year," doesn't includes the names of the victims, only what state the death occurred in, the location and the cause, and there are many causes. People die of heart attacks, vehicle accidents, helicopter crashes, snags (trees that fall on people due to fire damage), entrapments and, finally, they die in the flames. These deaths are called burnovers, and although "burnover" is the word listed in the column on the chart, it in no way conveys the horror of the event.

The Commonwealth Scientific and Industrial Research Organisation (CSIRO) is the federal government agency responsible for scientific research in Australia. In the agency's book, *Grassfires: Fuel, Weather and Fire Behaviour*, experts say that fire's maximum speed is between sixteen and twenty kilometers per hour (nine to twelve and a half miles per hour).[2] While the fastest human being can run twenty-three miles per hour, that doesn't mean you can outrun a fire. Last year, on the Spokane Tribes Reservation, I personally came across a herd of wild horses burned over in a field of grass that couldn't have been more than knee high. Fire surrounds, smoke asphyxiates, and if you don't have an escape route, you will lose. Animals, including humans, may be up to 75 percent water, but make no mistake, we are fuel to a fire.

No matter how well we believe we understand fire, it will always have a mysterious quality to it. We can break it down to its simplest parts: heat, fuel, oxygen and chemical reaction (the fire tetrahedron is taught to every first-year firefighter and revisited every year during recertification). We can study the weather patterns and topography of the terrain, test the air for humidity and have years of experience under our belts, but every wildland fire will undoubtedly do something you've never seen before. Fire acts more like a living thing than any other natural disaster.

In this book, we will investigate the reasons that the men and women who fight wildland fires in Oregon love what they do. We will also explore the land we all love so much and go over the fires that have occurred in our favorite Oregon areas throughout history. According to the U.S. Department of Agriculture, Alaska beats Oregon in acreage, but Oregon has more trees than any other state. There is something magical about our state and the relationship between the people who live here and our forests. The connection is hard to describe, but it's there. A study by United Van Lines has shown that Oregon has been the most moved-to state for the past four years, and there is little doubt our forests are a big reason for that.

Growing up in the Cascade Mountains, I come from a family of wildland fighters and loggers. I first started fighting wildfires in Hawaii while stationed at the army base Schofield Barracks on Oahu in 1993. I didn't come back to it until 2001, when I was in the Oregon National Guard. My unit was deployed to help out with the Biscuit Fire in 2002. While in the Guard, I went on to serve in Iraq and with Hurricane Katrina rescue operations in New Orleans, but after leaving the military for good, I started fighting fires again on a wildland engine with a private contracting company. There are many similarities between firefighting and the military.

I've been fighting wildland fires during the summer for years, and if you were to ask me what I love about it, I might tell you about the camaraderie on the fire crews or how much I love the adrenaline rush of danger, or I might go on about the sheer beauty of being off on some logging road untraveled for decades and looking down on a river winding its way through a national forest. I might tell you all of this—and it would be true—but there is something else I haven't pinpointed, and that sentiment is shared by everyone I've spoken to so far. While I can't say exactly what this unspoken realization is, I can say that it has to do with the pride of doing something worthwhile, protecting life and property and being a part of history.

Chapter 1

MAN VERSUS NATURE

On July 13, 2002, a dark cloud seeped over all of Southern Oregon, filling the sky for hundreds of miles with the black of a deep bruise. These opaque clouds pulsed with electricity until they opened up in a lightning storm over the Siskiyou Mountains with a natural violence unseen for 100 years or more. If primitive man were to have seen the succession of close to five hundred lightning strikes, it may have strengthened the belief in an angry God. Over two hundred of those lightning strikes turned into individual fires, and within a few days, they would grow into the Biscuit Fire, the largest fire the Pacific Northwest had seen in 137 years.

In 2002, the fire season started early in states like New Mexico, California and Colorado, and this pulled firefighting resources from the Northwest. Most of Oregon's firefighters were deployed to other states while the Biscuit Fire burned through the wilderness area and jumped the Illinois River, threatening homes and commercial forest land. The fire blackened nearly 500,000 acres, with some parts burning up to six hundred degrees, and it went on to burn for four whole months. The fire cost more than any other fire before it, with a final bill coming in at approximately $150 million.

The Biscuit Fire, according to Tom Spies, research forester at the U.S. Forest Service (USFS) Pacific Northwest Research Station, "burned in a wilderness area and ecological reserves that were designed to protect old growth and [northern spotted] owls."[3] The Klamath area had been burned only years before and teemed with new vegetation. This, mixed with the old growth, created a variable canopy. This means the fire on the ground

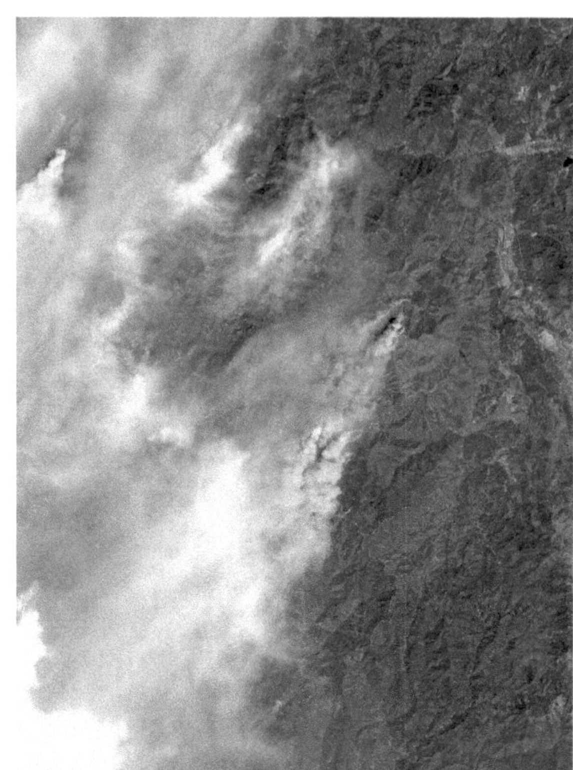

Right: NASA satellite image of the Biscuit Fire. *Courtesy of NASA.*

Below: Sometimes winds get so strong that they create a "firenado." This one was photographed by Marvin Vetter. *Courtesy of the Oregon Department of Forestry.*

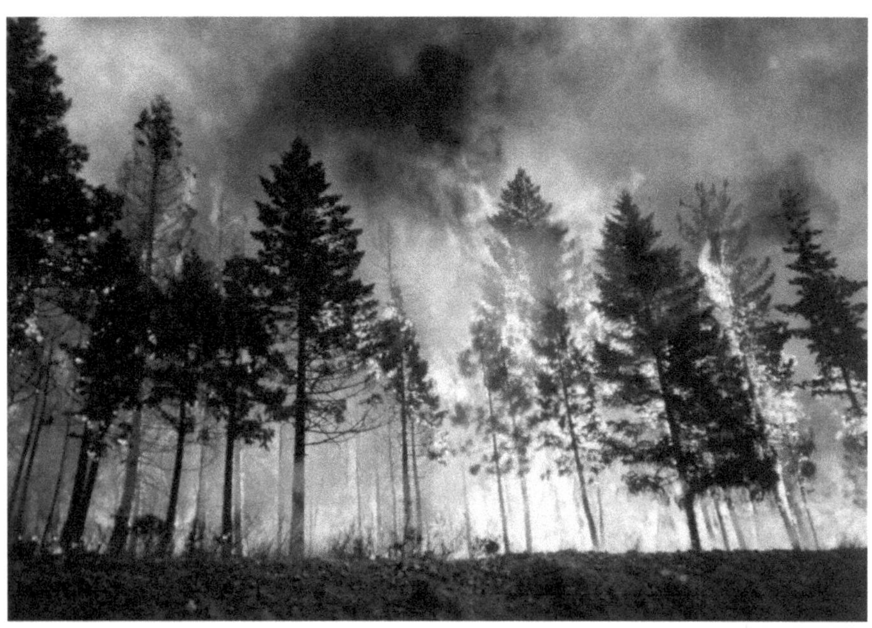

can use the different heights of the fuel to get into the canopy of the trees and start what wildland firefighters call a crown fire. Everything that burns is considered fuel, and when you have new and old vegetation, this creates fuels that can easily catch and burn quickly alongside fuels that take longer to catch but also burn longer. This mixed fuel classification is perfect for starting large and very intense fires. Think about how a person uses paper, kindling and logs to start a fire in a wood stove.

The smoke column sent up from the Biscuit Fire could be seen from space, and the fire spread, becoming the size of five hundred football fields. Walls of fire pushed through the forest like a tidal wave and turned everything in its path to ash. Flames of this fire—and those as big—create their own weather, and these weather patterns spit flames to create spot fires and wind that acts as a bellows to spread fire faster. Under the right conditions, a small spark can devastate hundreds of thousands of acres of forest by drowning it in an ocean of fire. Dust devils and even tornadoes made of fire, or "firenadoes," spin furiously in and around the fires.

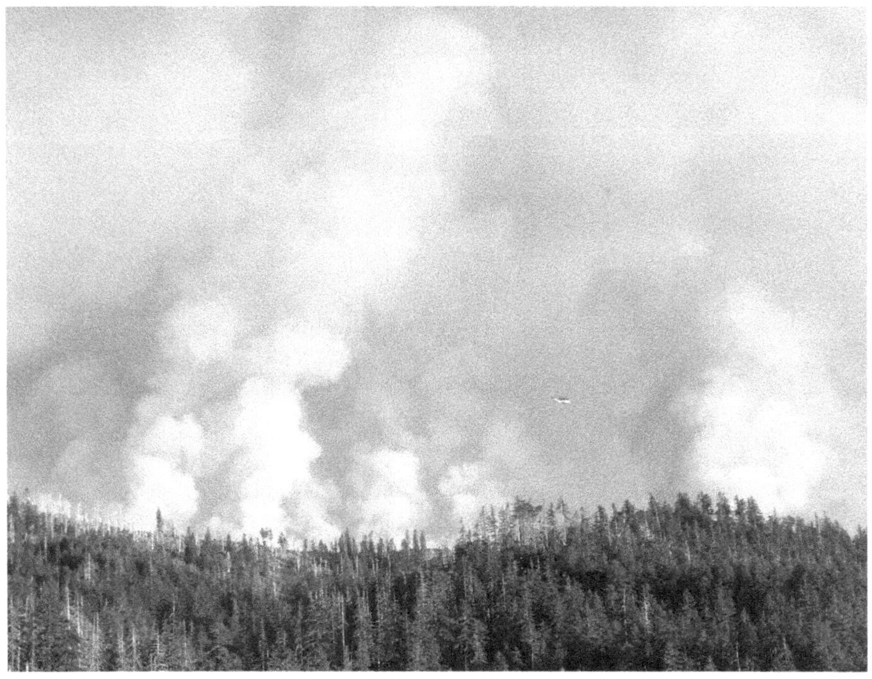

A wall of flames climbs the ridge on the Klondike Fire, 2018. *Photograph by the author.*

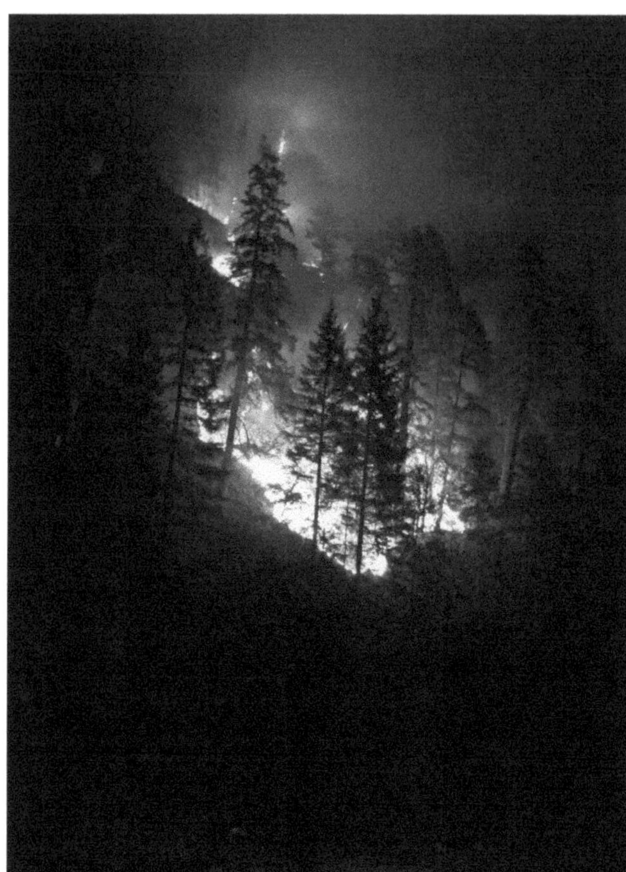

Left: Fire burning through the underbrush at night on the Flounce Fire, 2017. *Photograph by the author.*

Below: Oregon crew on Spokane Reservation. This was one of thirteen structures burned to ash. *Photograph by the author.*

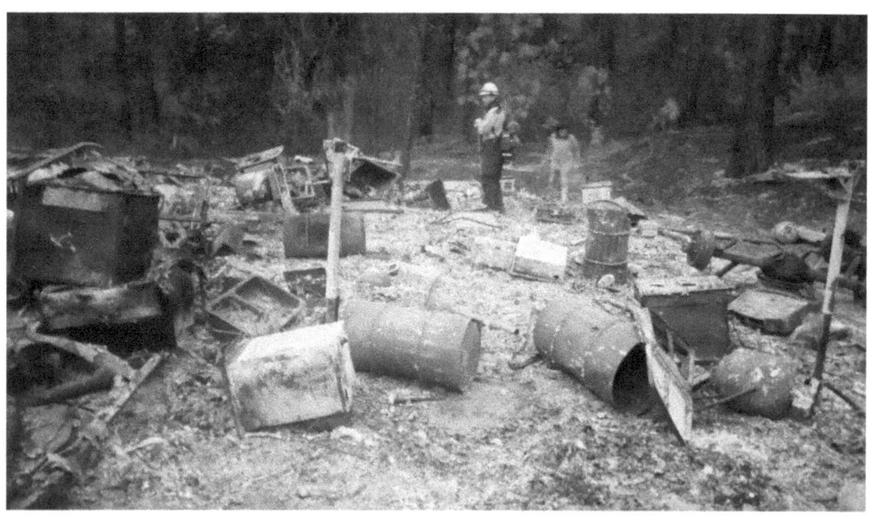

Everyone learns at a young age that fire is dangerous, but many don't really know the exact science behind burning. Burning is a chemical reaction between the oxygen in the atmosphere and a type of fuel. Fuel, in this book, means whatever burns: wood, shrubs and whatever is made of carbon (including people, unfortunately). At about three hundred degrees Fahrenheit, heat decomposes the cellulose, an insoluble substance inside the cells of plants, in wood, turning it to volatile gas (smoke). The material that is left is char or blackened wood. If the fire is hot enough, it will burn all carbon, leaving behind a white ash. This white ash consists of all the elements in the wood that will not burn (potassium, calcium, etc.). Wildland firefighters call an area of white ash "nuked" or "nuked out."

What a wildfire does to the land really depends on the intensity of the fire and the type of fuel. Most ponderosa pines and Douglas firs will survive when they are 50 percent or less scorched, but a spruce, lodgepole pine or subalpine fir will not. In fact, ponderosa pines may survive even if they are completely scorched, while other trees will not. This said, fire is not all bad for forests. In fact, the forest and wilderness in the Northwest had wildfires long before people lived in it. Wildland fires are a basic part of the natural process.

Fire returns nutrients to the soil much more quickly than waiting for logs to decompose. A fast-moving fire will burn away the underbrush and create open areas like meadows or allow new growth in a grove for deer, elk, rabbits, raccoons and other animals to eat.

Lodgepole pines actually need heat from wildland fires in order for their pinecones to open and release their seeds. In Eastern Oregon, regrowth of trees and plants wouldn't occur without a fire to burn away underbrush. Wildland fire is not always bad, but it is always dangerous, especially for towns and cities close to forested areas, which is the case in most of the state.

Oregon has been no stranger to some of the worst wildland fires in history. On a list of the all-time worst fires in North America, which would include the Chicago Fire and the Peshtigo Fire in Wisconsin that killed 1,700 people, Oregon would be on that list multiple times. The Great Fire of 1845 burnt from the city of Corvallis, in the Willamette Valley, to the Pacific Ocean. The Yaquina Fire of 1853, which happened six years before Oregon became a state, burned nearly a half a million acres.

Alaska is seven times larger than Oregon and has the most forested area of any state, but Oregon comes in second. According to the Oregon Department of Forestry, "Oregon's forests cover more than 30 million

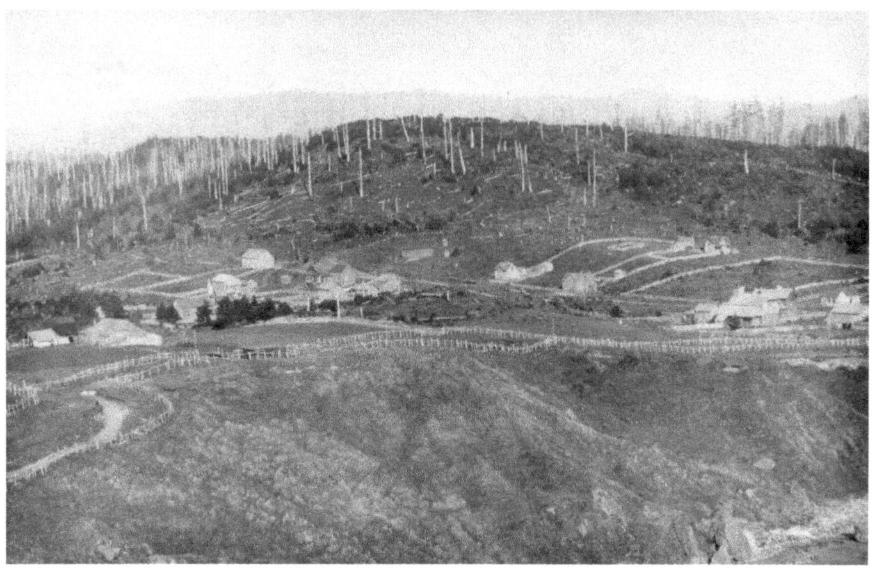

The Great Fire of 1868 changed the face of the state. Here is a photograph of Port Orford in 1902 showing a landscape still barren except for burned and dead trees. *Courtesy of Port Orford Historical Photos (Alan Mitchell Collection).*

of the state's 63 million-acre land base, or about 48 percent of the state's total landmass."[4] What do we do when a fire rips through our forests and endangers our homes and businesses? How do we fight this incredibly violent force of nature? We call our firefighters.

Over the last fifty to sixty years, firefighters have added machinery like helicopters and bulldozers to fire prevention and suppression efforts, but how they fight the fires has stayed pretty much the same—a group of men and women with tools in their hands. The bulk of the firefighting happens by sawing trees and removing fuel with saws and by manually scraping the ground to mineral soil in three- to five-foot-wide paths with hand tools.

Over the years, the PPE (personal protective equipment) has improved, but we've always fought fire in a "man versus nature" way. The best tool a firefighter has is his or her brain, but he or she will also need a strong back and endurance. A good firefighter tries hard to predict fire behavior by taking in the weather, the wind direction and speed, the terrain and many other aspects, and they're also asked to perform incredibly hard twelve- to sixteen-hour days for two weeks straight without getting a day off.

In this book, fires outside of urban areas will be called wildland fires and not forest fires. The reason for this is that not all non-urban fires in Oregon

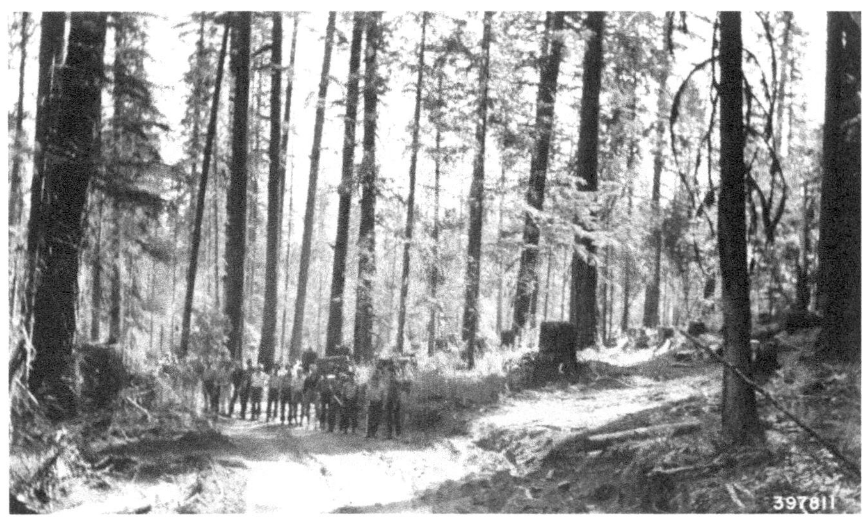

The "Willamette Flying Twenty" special firefighting crew lined up in position in Lane County in the Willamette National Forest, 1940. They were getting ready to dig some line. *Photograph by Roy A. Elliott, courtesy of the Gerald W. Williams Collection at Oregon State University.*

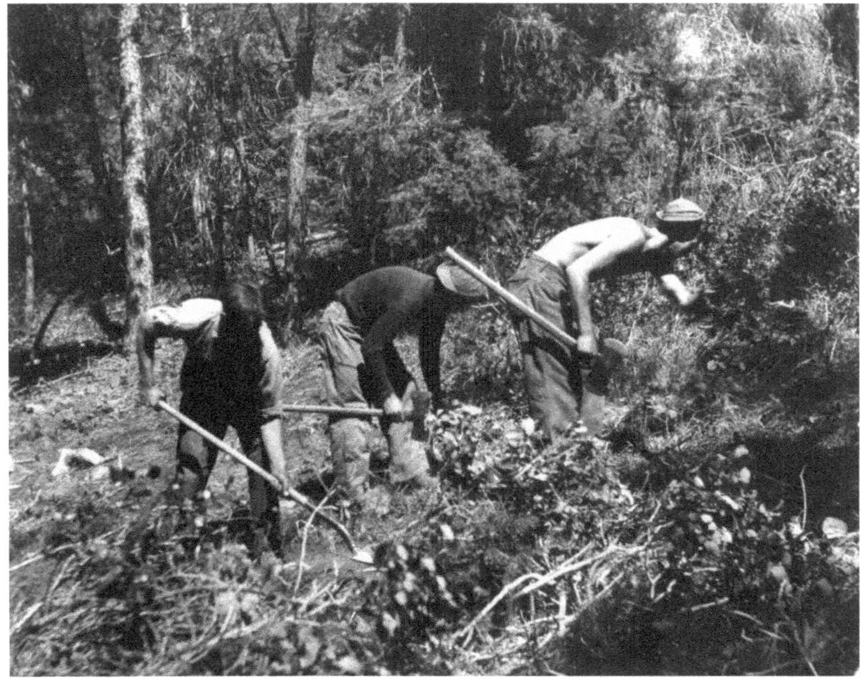

Wildland firefighters William Baller, Merle Haskins and Tom Ballantine digging a fire line during training in the Ochoco National Forest, 1933. Photograph by C.M. Rector. *Courtesy of the Gerald W. Williams Collection at Oregon State University.*

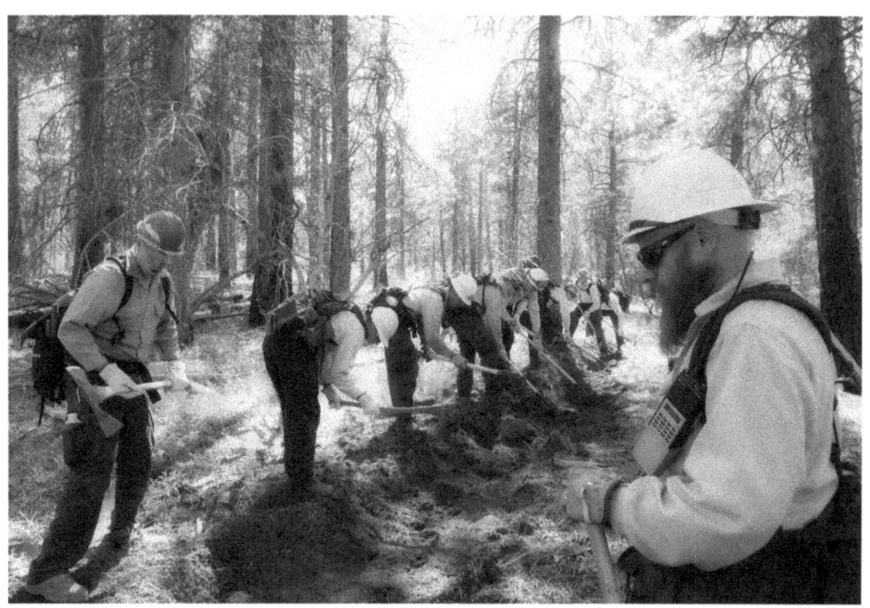

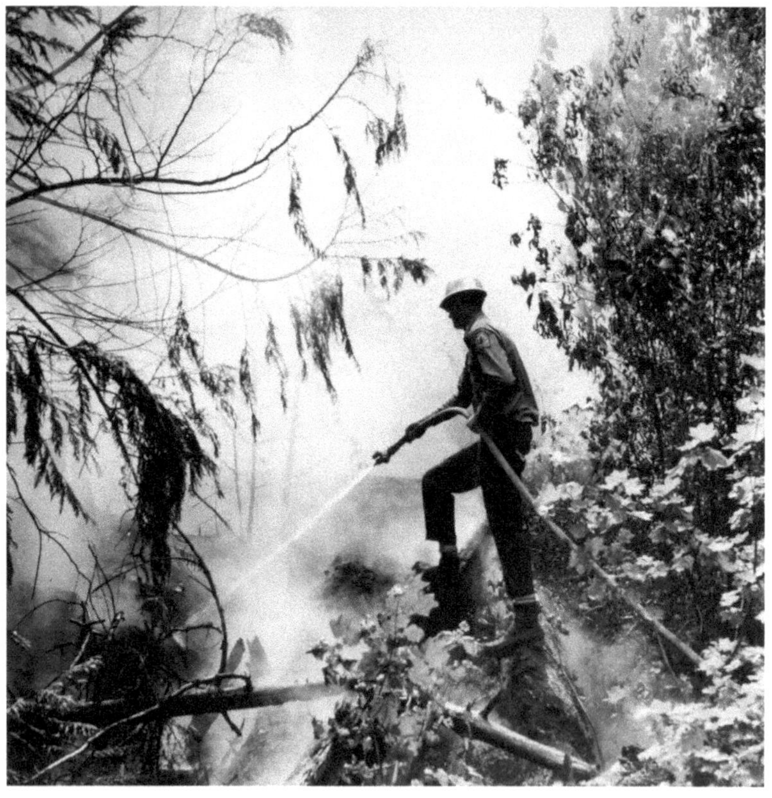

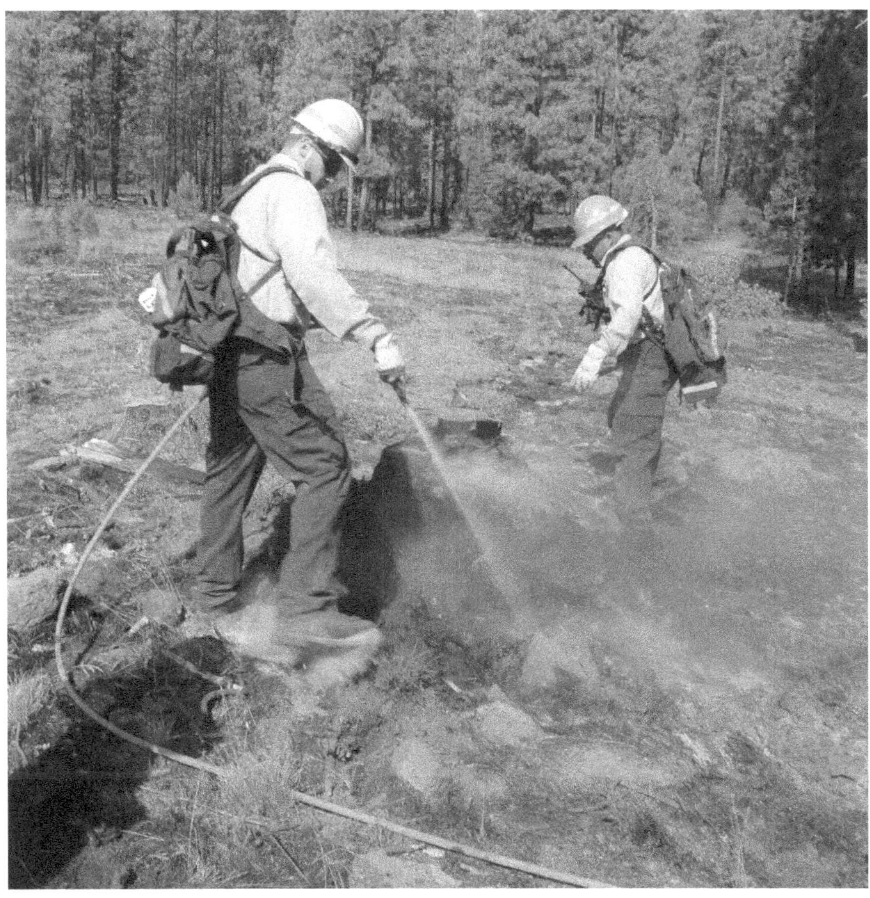

Above: U.S. Forest Service wildland firefighters spraying a stump fire on the Bryant Fire, 2014. *Courtesy of the U.S. Forest Service.*

Opposite, top: New firefighters being trained at Carmen Thomas Memorial Guard School to fight wildland fires in case they are needed in the upcoming fire season, 2013. *Courtesy of Wikimedia Commons.*

Opposite, bottom: Phillip Voss of the Oakridge Ranger District uses a hose to put out a blaze that happened on private land within the zone of responsibility of the Willamette National Forest, August 1967. This fire was set by children playing with matches. *Photograph by Samuel T. Frear.*

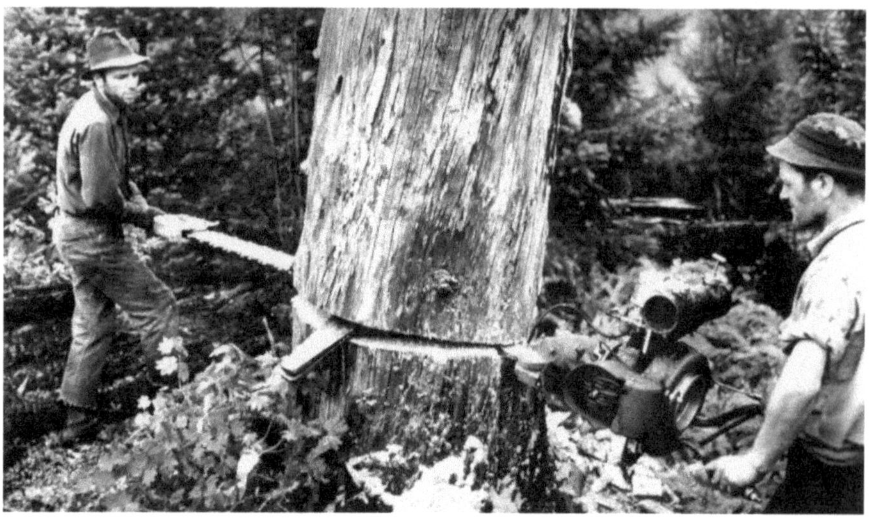

This photograph shows the very first use of a gasoline-driven power saw, or chainsaw, in the field in the United States. The first tree to fall to a chainsaw was in the Tumble Creek Fire near Detroit Lake in the Willamette National Forest in 1941. *Courtesy of Forest History Center/ Oregon Department of Forestry.*

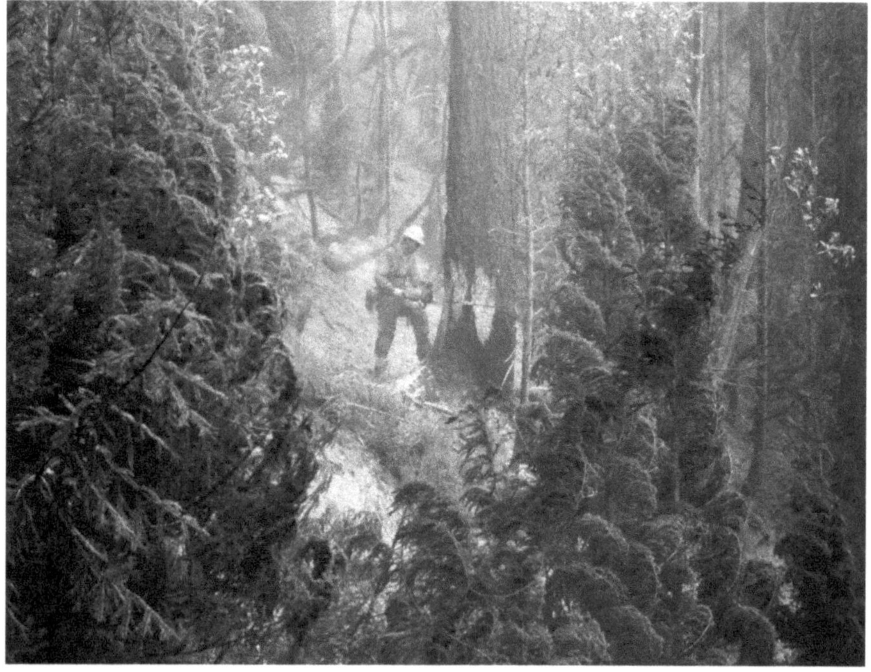

U.S. Forest Service sawyer falling a snag in the Brimstone Fire in 2013. *Courtesy of the U.S. Forest Service.*

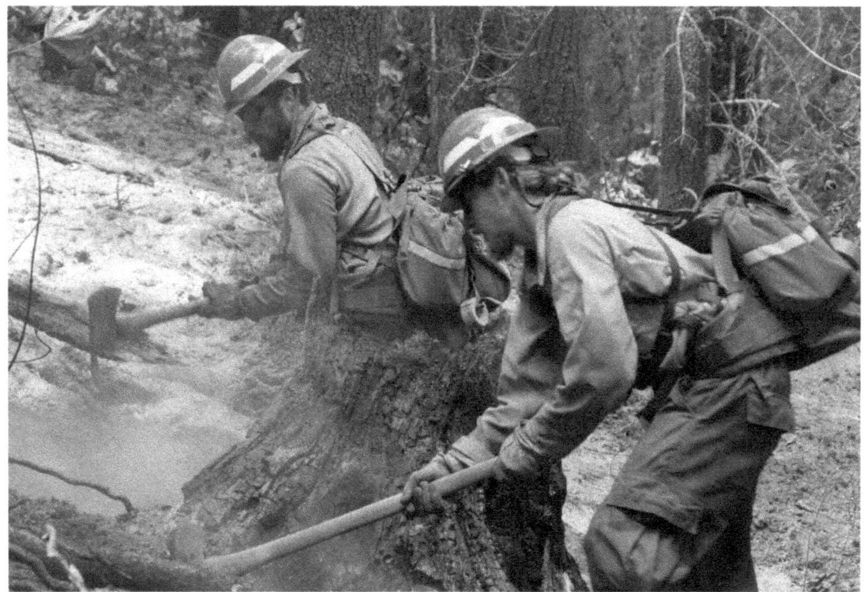

Two wildland firefighters digging out and dry mopping a stump fire. *Courtesy of the U.S. Forest Service.*

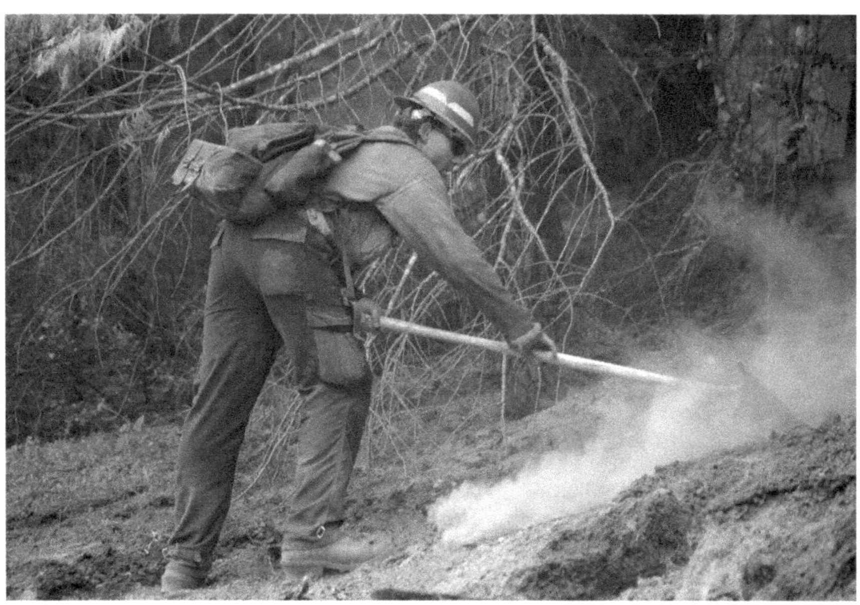

A wildland firefighter works hard digging out a root fire at the Horse Prairie Fire, 2017. *Courtesy of the U.S. Forest Service.*

burn in the forest. In fact, only 48 percent of Oregon is forested. West of the Cascades, eight out of ten trees are Douglas fir, making it the state's most common tree, but east of the Cascades, ponderosa pines hold the title. Eastern Oregon has the high desert, and western juniper and mountain hemlock can be just as dangerous when on fire. Grass fires, for that matter, often travel the fastest because they ignite so quickly and can move at the speed of the wind.

When it comes to man versus nature, to win, man must understand nature. Whether the fire is a Type 5 incident with one engine or hand crew fighting it or a Type 1 incident with a thousand personnel, weather should be taken every hour. You can get an app for a tablet or smartphone to do this, or use a handheld digital weather meter, but every crew that fights a fire will have a belt weather kit and know how to use it.

With the belt weather kit and the Incident Response Pocket Guide, any wildland firefighter can predict the weather with some accuracy. Smartphones and devices that provide weather information can be more accurate, but they can also, many times, be less accurate. Also, crews will often find themselves fighting fire in a zone without cell coverage or satellite connection, so electronic devices won't work anyway.

Belt weather kits should include several items like the sling psychrometer, an anemometer, a small bottle of distilled water, a compass and more. With a kit, a firefighter can get the relative humidity (RH) in the air, the speed and direction of the wind, the dew point, the probability of ignition (POI) and more vital information needed to predict fire behavior. In terms of RH, for instance, the higher the relative humidity (water content) in the air, the lower and slower a fire will spread and the lower the probability of ignition. The wind is another very important factor to take into account before and while fighting a fire. Not only does the wind deliver the oxygen needed for the fire to burn, it also gives it a means of moving fast, and the wind does just that on the coast and in the high desert and areas like the Columbia Gorge in Oregon.

In many parts of Oregon, the wind will consistently blow one way in the morning and then completely change direction in the afternoon. Not only does the wind bring oxygen and move the fire, it brings other dangers as well. It can bring dust and smoke, which limit visibility; it can break off burnt, rotted or broken branches that can fall on wildland firefighters (these are known as widow-makers); or an entire tree can fall (these are called snags) and shoot embers in front of the head of the fire across the fire line, roads, creeks, and so on (spot fires).

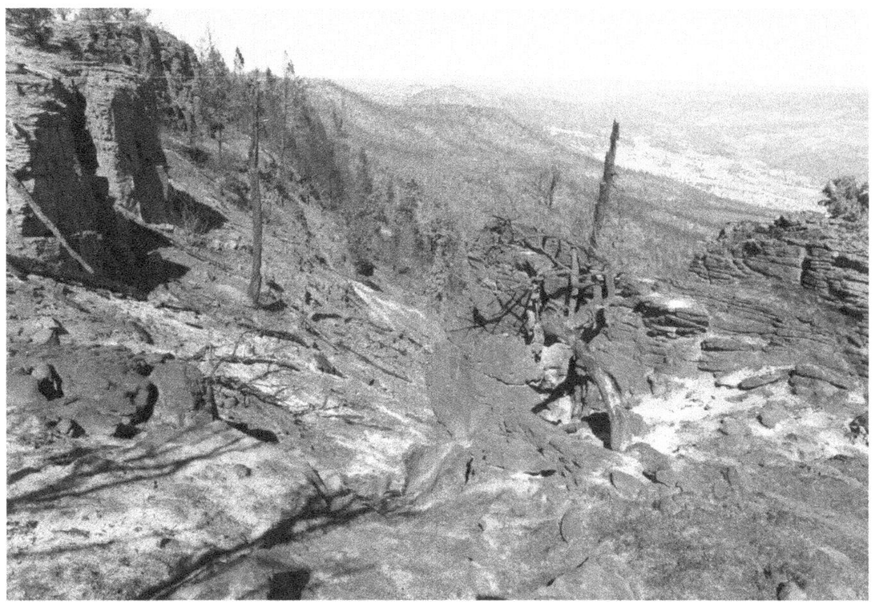

The Lost Hubcap Fire of 2014 burned almost 2,800 acres near Monument, Oregon. Around 70 percent of the fire was timber and underbrush, and the other 30 percent was grass and sagebrush. *Courtesy of the Oregon Department of Forestry.*

The absence of wind can also cause big problems within a fire. Without wind, the smoke and clouds hover over a fire, creating an inversion. This keeps pollutants from dissolving into the atmosphere, makes the air unhealthy for wildland firefighters to breathe and reduces visibility. The stagnant smoke kept in place by inversions can also heighten the anxiety of residents in the area as well as wildland firefighters. Not being able to see the sun or sky—or even a dozen feet in front of you—for a week or two straight can reduce morale in surprising way.

Smoke is made up of small particles, gas and water vapor. The U.S. Fire Service says the gas consists of "carbon monoxide, carbon dioxide, nitrogen oxide, irritant volatile organic compounds, air toxics and very small particles."[5] Carbon monoxide becomes unhealthy over 70 parts per million. At 150 parts per million, a wildland firefighter will experience headaches, dizziness, fatigue and shortness of breath, and when he or she is under an inversion, there is no place to go for fresh air. It may feel like he or she is slowly suffocating. The only thing to do, other than being medically evacuated, is to put a scarf, rag or shemagh over the mouth and keep working.

Of course, wind and weather aren't the only factors one needs to consider when fighting fire. The more seasons you have under your belt, the more you learn. Terrain and types of fuel (vegetation, structures, anything made of carbon, et cetera) are also very important. As any Oregonian knows, we have a great variety of both terrains and fuels. Grass fires in a field will move fast, but a grass fire on a slope will move faster, and grass growing up a box canyon will move the fastest because the terrain acts like a chimney, funneling the fire. Fire will spread faster on a slope that faces the rising sun because the sun will lower the relative humidity and dry the fuel.

Fighting fire is a science, and every single person who fights it, if they are any good, is a scientist in their own way. The wildland firefighters who stay season after season need to know all of this as well as how to use all the equipment. Part of the continuing education of a wildland firefighter is to look at the intensity of a fire and the terrain, predict the weather, know the fuel types, find wind speed and direction, know things about the area that only a local would know and much more. Then, he or she must put all this together and call for the correct resources.

Not only is this a career for a person with physical strength and endurance, this is a job for someone who can make all these calculations. A person can get away with only doing what their supervisor tells them for a long time, but sooner or later, everyone gets put into a position where they will have to use what they've learned and be pushed to their physical limit. Firefighting is not a job for everyone, but if you can do it, there is no more rewarding career.

Chapter 2

THE AGENCIES AND ORGANIZATIONS

There are many different organizations that fight fire in Oregon, and they all have different philosophies and priorities when responding to a wildfire. Much of the time, the fires burning in Oregon are being put out by people who don't live in the state. Every state has several wildland firefighting crews of all types that are deployed over state lines for an incident. At any big fire in Oregon, in any year, there can be fire crews from every corner of the country.

This means a complex fire (one fire consisting of smaller fires that usually burn into one) in Oregon can have a leadership team from New York, heavy machinery from Colorado, hotshot teams from Washington, hand crews from North Carolina and possibly only a few Oregonians. The resources are called up and sent to incidents as they happen, and everyone works together no matter where they are from. This is why everyone who works on a wildland fire, or any incident that may come out of a wildland fire, must complete online FEMA training like ICS-100 and ICS-200. This includes federal, state, territorial, tribal, local and private sector nongovernmental personnel. ICS stands for "incident command system." It began in the 1970s after a large series of fires in California. The purpose of the ICS is to help all the various types of personnel on an incident work together as seamlessly as possible, giving them all standard operating procedures and a strong chain of command.

By far, the biggest agency tasked with fighting wildfires is the U.S. Forest Service (USFS).[6] The USFS is an agency of the U.S. Department of Agriculture and manages and protects the national forests in forty-three states and Puerto Rico. That means it is responsible for roughly 193 million acres of land across the country. The U.S. Forest Service's motto is "caring for the land and serving people," and it does that by managing public lands and giving technical support and financial assistance to state and private forest agencies.

The USFS has one of the longest-running public service advertising campaign mascots in history: Smokey the Bear. Smokey has been warning people about the dangers of wildland fires for over seventy years. When people think of the USFS, wildland firefighting crews may come to mind, but the USFS also employs teams of scientists who help maintain the forests and keep them healthy for future generations. Thirteen of these scientists were awarded Nobel Peace Prizes in 2007 for their work with the Intergovernmental Panel on Climate Change (IPCC). The USFS is the largest and most technically advanced forestry research agency on the planet. The scientific teams study topics such as climate change, how to stop invasive plant species, saving endangered wildlife and much more. During the Eagle Creek Fire in the Columbia River Gorge National Scenic Area, the USFS sent archaeologists to assess risks to high-value cultural resources and historic sites in order to prevent anyone from getting hurt after the fire was extinguished.

These scientific teams are amazing, but firefighting still boils down to the men and women on the ground. As of 2018, over ten thousand professional wildland firefighters were employed by the USFS.

Oregon is made up of 68,018,240 acres of land, and about half of that is owned by the federal government. Twenty-three percent is managed by the Forest Service, and another twenty-three percent is managed by the Bureau of Land Management (BLM)—this equals about 15.7 million acres apiece. The Department of Agriculture is the mother agency of the USFS, but many of the other federal agencies that fight forest fires fall under the Department of the Interior, including BLM, the National Park Service and the Bureau of Indian Affairs. All of these groups have an interagency agreement with the Oregon Department of Forestry (ODF) as well as city and county fire departments. In an interview with Jim Gersbach, the public information officer for the Oregon Department of Forestry, he said,

> *Oregon approaches wildfire management as a single coordinated system. Different agencies have specific responsibility for certain lands, but all cooperate with the same goal in mind—to suppress wildfires so they don't*

spread uncontrollably. Most areas of the state operate under the "closest forces" concept, whereby the nearest resources respond immediately to provide the initial attack on a fire. The jurisdictional agency retains overall responsibility for the fire and usually assumes management of fires that escape initial attack. The way this works in practice generally is that if a wildfire originates in a National Forest it is the responsibility of the U.S. Forest Service to suppress that fire. If it originates on land protected by ODF then it is our agency's responsibility. However, a lot of wildfires burn across multiple jurisdictions, especially because of Oregon's checkerboard pattern of land ownership (in some places private and federal land ownership alternates every mile). In such cases, an agreement will be reached between the agencies over who should manage that wildfire.[7]

In the 1960s, 1970s and early 1980s, many small towns and counties with large forests had at least seasonal teams that would thin forests during the off-season and fight fires during the summer. A lot of these crews were at least partially funded by brush disposal funds. Brush disposal funds came from big timber companies like Warehouser, Stimson Lumber and others. These companies had to pay a fee for brush disposal for every one thousand board feet of lumber they harvested. When the timber agency declined, so did the brush disposal funds, and as a result, the small towns and country crews went away.

The state and federal governments are under constant pressure to reduce their budgets. Most of the fire crews, whether they are federal, state or contractors, are paid with tax money, and the majority of the firefighters hold seasonal positions. Many politicians don't feel that wildland fire prevention is a priority. Unfortunately, many times more tax money goes to reacting to the fires of the season than would have been spent on the prevention of the fires. For a good example, look at the 2015 and 2017 fiscal years: 2017 was the most expensive year on record for wildland fires, yet the Trump administration proposed cutting $300 million from USFS wildland firefighting initiatives and $50 million from its existing wildland fire prevention programs. It also proposed reducing volunteer fire departments' federal funds by 23 percent around the nation.[8]

Residents in some areas of Oregon have elected to not pay taxes for fire services. The people who live in rural areas that have no forest, where fires are rare, elected to pass legislation to create their own fire response and not be taxed, so the Oregon Department of Forestry and the USFS don't offer fire response coverage in some place like parts of Douglas and Josephine Counties.

Some of the small towns in these areas have fire departments for structures and homes but don't have the equipment for wildland fires. Mid-Columbia Fire & Rescue chief Robert F. Palmer says that many of the ranchers pulled together and provide their own fire protection, and that works for them. The people working the land band together and respond to their own fires, or landowners will hire contractors to respond for them.

There are other agencies that fight fire as well. The government-funded Civilian Conservation Corps (CCC) is long gone, but many states have similar programs run with money from donations and grants. Many fires also utilize convict crews. They can dig lines, do laundry, cook and serve food and other such jobs. Veterans are discovering that wildland firefighting is very much like all the parts of the military a person enjoys but without the combat. Team Rubicon, a national veterans' group that responds to natural disasters across the country, has started to help veterans become trained and get certified to fight wildland fires. Team Rubicon has even deployed crews to fires over the past few years. National Guard units are also trained and sent out to fight fires and control traffic during incidents around the state.

Chapter 3

THE MEN AND WOMEN WHO FIGHT WILDLAND FIRES

While the federal, state, county and city organizations that fight wildland fires today have state-of-the-art equipment, fires will not go out without the work of men and women with tools in their hands. U.S. Forest Service air operations branch director Sam Swetland, who spent forty-two seasons fighting wildland fires, said it best: "Fires don't go out without boots on the ground....Aircraft don't put it out, water doesn't put it out—just people swinging tools."[9]

Swetland started on a hand crew in 1976. He had a four-year degree in biology from the University of Oregon, but in the mid-1970s, the job market wasn't the best. He decided to take a seasonal job with the U.S. Forest Service (USFS) at the ranger station in Blue River, Oregon, a small town off the McKenzie River in the Willamette National Forest. After one season, he was hooked, and he worked his way up to air tactical group supervisor and air support group supervisor. He officially retired from the forest service in 2008, but he is hired back whenever the USFS needs his skill set, which happens almost every year.

When asked why he did the job so long and why he keeps coming back, why he stayed in the woods working some of the hardest labor a person can perform for a job when he could have gotten an easier job with his degree, he told me it was the hard work in the woods that hooked him. There is something near indescribable about working twelve to sixteen hours of hard labor each day with a crew. "It's the teamwork factor. It's learning leadership skills on something that matters."

In 2017, Sam Swetland was called on again, but this time on a fire close to home. The Rebel Fire burned only 8,703 acres, but it was only a mile from Swetland's house along the South Fork of the McKenzie River in the Three Sisters Wilderness Area. He suited up and helped get that fire under control.

Keith Hergert started West Coast Wildland Strike Teams with Barbara Kuykendall. They are a small but growing contract company that has been fighting wildland fire since 2000. Hergert said that he started his company because the forest service teams were shrinking. When the timber dollars started to dry up, fewer and fewer people were hired to be on hand crews and engines. Simply put, he saw a need. Today, West Coast Wildland Strike Teams has around fifteen employees and four engines.

It's not easy to run a business and compete with larger companies like Grayback, PatRick and others that have fleets of vehicles and hundreds of employees. At the end of every year, Hergert has to inventory his vehicles and equipment and replace anything that needs to be replaced and perform expensive maintenance on all his vehicles. Driving in the mountains all summer, often on dozer lines and unimproved dirt roads, takes a toll. He also has to go to the annual contractor get together in February to meet with representatives from the USFS, Oregon Department of Forestry, Bureau of Land Management, Department of Natural Resources (from Washington State) and other government agencies to find out if there have been any changes in the rules and regulations for fighting fire in the region. Most years, there are changes, and if Hergert wants his company working on fires during the summer, he needs to implement these changes before fire season starts. Sometimes these changes can be very expensive, especially if they have to do with updating communication systems or buying better personal protective equipment.

Another difficulty for West Coast Wildland Strike Teams is staffing their engines. This is a seasonal job, and keeping people year after year is very tough. They do their best to hire veterans and men and women from local fire departments who would like to go from their structure fire jobs to fighting wildland fires. Many years, they try to hire young people right out of high school, but working twelve to sixteen hours a day for two weeks straight is more than some can handle.

Once a contracting company meets all the rules and regulations, has maintained all its vehicles and equipment and has a full roster of available and qualified candidates, the owners wait. The company goes on a dispatch list. The order of this list is decided by how long a company has been in business, the evaluations a company has received from first line supervisors while on fires and how much the company charges per day for the use of its vehicles (this includes pay for the company's firefighters).

This year, because of a single employee, West Coast Wildland Strike Teams had to turn down four dispatches. This is four points against the company, and because of that one employee, Hergert's company and all his employees could suffer. But that doesn't get Hergert down. He says, "That's the cost of doing business. If it were easy, everyone would be doing it."

Hergert and Kuykendall run a family-oriented business, and even though Hergert is sixty-six years old, he's still out there with his crews fighting fires. He started this company so he could be out there on the front lines, and he will do it as long as he can because it's what he loves to do. He has been fighting wildland fire for eighteen years, but he started volunteering at his local firehouse in Sandy in 1973.

Tyee Burwell was born into the timber industry. His father, Bill, and grandfather David Burwell were in the timber industry, and Tyee and his brother Lakota were loggers, too, at least for a while. From the mid-1980s until the mid-1990s, the demand for Oregon timber declined all around the world. That, combined with environmental groups' protection of the spotted owl's territory, devastated small logging companies and the communities that depended on the logging dollars. Many of the families who had been loggers for generations transitioned into working for the USFS or the Oregon Department of Forestry or started their own contracting businesses. Tyee's father, Bill Burwell, started his own contractor business with a partner, and that's how Tyee and his brother Lakota began their long careers as wildland firefighters.

Tyee's skill with a chainsaw got him onto the Rogue River Interagency Hotshots. He was administratively hired and loved working with the group, but the drive from home was a bit too much. After six months or so, he applied and was accepted to work with the Zigzag Interagency Hotshots on Mount Hood.

During our interview, Tyee Burwell talked about a time before he joined any hotshot crew. His scariest moment was when his crew was nearly killed by a burnover. The hand crew had been called from Oregon to fight the Clear Creek Fire in Idaho in 2000. They'd been chasing lightning strikes in the Frank Church River of No Return Wilderness. That year, eleven of the twenty wildland firefighters in his crew were first-year rookies who had never deployed to a fire. Tyee was the assistant crew boss, and half of his crew were on their very first fire. The Incident Action Plan (IAP) and their morning briefing let them know the size of the main fire above them and how it would move that day. They also knew the Texas Canyon Interagency Hotshot Crew below them would be blackening the line—this is how wildland firefighters fight fire with fire. Blackening the line is the practice of using a fusee or drip torch to burn away the underbrush of an area in order to remove fuels so that when the larger fire gets to that point, it has nothing to burn, and this way, it could be contained or at least controlled.

The hotshots to their south started to burn their designated area, but during the mission, one of them suffered a life-threatening spinal injury. The crew had started the burn operation but left it because they went into crisis mode and needed to get medical attention for their injured crew member. Meanwhile, the fire they set kept burning.

The standard operating procedure for an critical injury at the time was to cease all radio traffic except for radio traffic concerned with the life-threatening injury. The fire updates stopped, the weather reports stopped and all the questions Tyee's crew had about the fires to the north and south of them were not answered.

Soon, the wind shifted, and the smaller fire from the unfinished and unsupervised burn operation to the south spread and made a run to the bigger fire to the north. This happens more often than people know. Different fires in the same area almost always burn together. The small fires burn toward the larger fire. Tyee and his crew boss recognized this was happening. The two fires were going to burn together with them in the middle. They needed to move, or they would get caught between the larger fire from the north and the smaller fire to the south.

Tyee's crew boss, Steve Clarke, who is now a professor at Oregon State University, had to think quickly. He pulled out his map and compass to search for a piece of land where the crew would be safe. The fire to the south had cut off their escape route, so they'd have to find another way. The safety zone they were assigned was unsatisfactory and wouldn't do much to protect the crew, so he had to find another option. After a few moments, he found a clearing a few kilometers away, but uphill. Going through all the information he had at the time, he felt there was no other option. They couldn't go where the IAP told them to go, and if they didn't make a run for it, the trees surrounding them would burn over in a matter of minutes.

Crew boss Steve Clarke shot an azimuth to the area he picked out, and he led the crew with his map and compass. Tyee Burwell's job as assistant crew boss was to bring up the rear of the formation so no one was left behind. The southern fire followed them so closely that Tyee could feel the heat

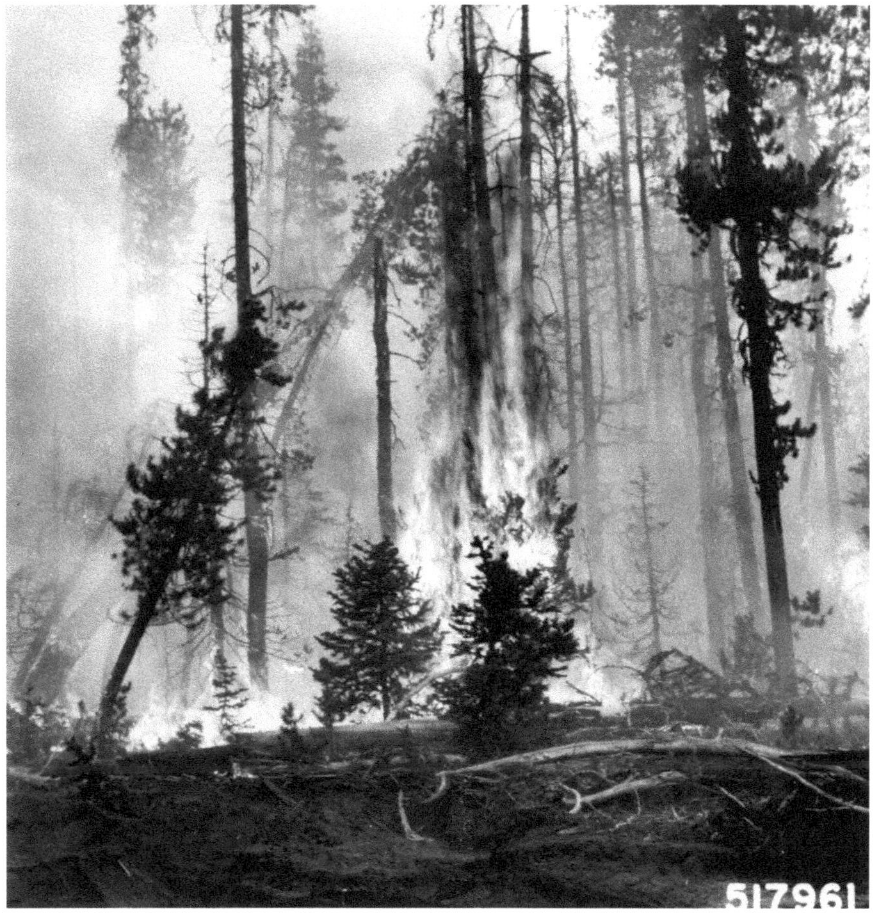

Intense flames roar through trees on the McKenzie Airstrip in the McKenzie Ranger District, August 1967. This is an example of a fire climbing trees into the canopy and crowning. *Photograph by S. Frear, courtesy of the U.S. Forest Service.*

through the back of his Nomex greens and yellows. Several times, one of his team members stopped and told him she couldn't go on. He couldn't carry her, and they'd need the tools she carried. The only thing he could do was keep his cool while others lost theirs. Looking back, he saw the fire had crowned, which means it climbed up into the canopy, the branches of the trees. The fire to their north raged above the treetops. Squirrels, raccoons and birds hurried out of the main fire as it made its way toward them. Some of them were on fire; the animals themselves inadvertently helped to spread the fire as fast as they could sprint.

A wall of fire one hundred feet high chased them like a crashing tidal wave, and the main fire came at them like an avalanche of heat and flame. The sap in many of the trees boiled and exploded. The fire sucked in the air so hard and fast in both directions that it felt like a tornado. Tyee had to shout instructions at the top of his lungs. One of his crew grabbed at her ankle, dry-heaved and said she couldn't go on. There was no stopping. If any of them did, they would surely die.

Crew boss Steve Clarke kept his azimuth, and Tyee Burwell kept everyone in front of him, and it may have felt like an eternity, but they made it. The crew burst into the clearing that was about three-quarters of the length and width of a football field. Right away, Clarke lit a fusee (a kind of flare) and burnt the tall grass. Grass fires burn faster than fires in trees. Soon, the field was barren of all fuel, and Burwell and all of his crew members were in the middle of the black. Flames spread through the old-growth canopy and surrounded the field, spitting fire, smoke and ash hundreds of feet in the air.

They were in an oven surrounded by a forest on fire. The noise was deafening, and the heat was so intense that Burwell remembers pressing his thumb down on the plastic of his helmet and leaving a print. If it wasn't for their fast thinking and land navigation skills, all twenty of the members of this hand crew would have died. The field they had burned provided enough space that they didn't have to deploy their shelters, but Burwell says they probably should have. Burwell and his crew never blamed anyone for negligence or letting the burn operation get away and almost frying them. "They were doing what we would have been doing, taking care of their own. It was no one's fault—just bad luck."

This event was a learning opportunity, and it spawned what we now call the "incident within an incident" protocol. When an accident or injury occurs on a fire, it is now given its own incident commander and resources, and they all communicate on a separate designated radio channel. As for Tyee Burwell, did this brush with death shake him or make him think about quitting? No, and it wasn't until after this experience that he became a sawyer on two different hotshot teams.

Chapter 4

WHAT YOU NEED TO KNOW ABOUT FIGHTING WILDLAND FIRES

The very first thing to know when it comes to fighting wildland fires is L.C.E.S. (**L**ookout, **C**ommunication, **E**scape routes and **S**afety zones).

Lookout(s) is/are wildland firefighters put in place to watch the fire, spin the weather, pay attention to trigger points and communicate all these things to the crews working on that fire. Every and any fire can behave in an unexpected way. Lookouts directly watch the fire, fire behavior and weather. Each incident dictates the number of lookouts needed. This depends on the size of the fire, the terrain and other factors decided by the person in charge. There should always be a lookout no matter the size of the crew or fire.

Communication is the how messages are delivered to wildland firefighters. The message may be an alert to a hazard, coordination with other crews or units or relaying important information. Even if a firefighter has a radio or cellphone or is within yelling distance of another person, communication must be reliable and timely.

Escape routes are the paths the firefighter can take from wherever they are when danger presents itself to a safe area. Many times this safe area is the safety zone. This danger can be seen firsthand or communicated by a lookout. Every firefighter should have at least two escape routes from where they are working to safety. The conditions and situations of the job he or she is tasked with may force a wildland firefighter to change escape routes at any time.

Safety zones are locations where a wildland firefighter can find refuge from any danger. This does not mean a deployment site. A safety zone is a place where a firefighter can fit an entire crew and all their equipment, and no one will need to deploy a shelter.

A person would be hard pressed to walk into any fire base and find a wildland firefighter who doesn't know what L.C.E.S. stands for. The concept has been around for a long time, but it became something that all wildland firefighters had to learn after former Zigzag Interagency Hotshot superintendent Paul Gleason wrote a paper about it in June 1991.

Today, when fire season starts, most of the people who fight wildland fires are seasonal workers. The seasonal firefighters start to oil their boots and pack their fire bags around mid-spring. Along with the items listed earlier, fire bags will have at least two pairs of greens and two yellows. Greens are cargo pants, and yellows are long-sleeved collared shirts. Greens and yellows are made of Nomex, a fire-resistant, synthetic meta-aramid fiber developed in the 1960s by DuPont. Everyone who works on a wildland fire must wear approved Nomex materials. Today, many different companies sell Nomex pants, and not all of them are green, but yellow shirts are mandatory.

Many seasonal wildland firefighters are surprised by the extraneous work and long hours. Because of the extreme physicality of the work and the toll it takes on the body, there is a lot of turnover, leaving plenty of opportunity for new people who have never tried it before. Every year, the USFS and the Oregon Department of Forestry hire new firefighters, many of whom are contractors. While most employers provide the greens, yellows, a shelter (which will be covered shortly) and maybe some gloves, they don't provide much else. If you want to try out firefighting, or know someone who does, here are some helpful facts.

Good boots should be a top priority, especially for hand crews who walk hundreds of miles in a single season. Good socks are also a must-have. A wildland firefighter can't have enough socks. Depending on your crew's job, you may change socks a few times a day. There is no self-contained breathing apparatus for individuals like in structure fire, so pack plenty of handkerchiefs or shemaghs to wrap around the face in order to keep out smoke. Another very important piece of gear is the pack that will hold the shelter. Every wildland firefighter in the United States must carry a fire shelter with them. You must have something in which to carry your shelter. An old military style Y harness with belt and ALICE clips will work, but while you're on the line, you will need to carry important things like water, ear protection, eye protection, fusees, a compass, gloves and more. A good line pack will do the trick, but remember, the bigger the pack you get, the more weight you'll end up carrying. This seems like a no-brainer, but a good medium-sized pack will force a wildland firefighter to take only the essentials. The more pockets you have, the more "must-haves" you will stuff in them.

The first thing a person should understand when considering whether to be a wildland firefighter is that this is one of the most physical jobs a person can choose. A hand crew will walk miles through the woods to get to a fire, and once they arrive at the fire, they will use hand tools to cut fire line. This means scraping a path into the forest floor to mineral soil that is at least three feet wide around the perimeter of a fire. This path will go through underbrush, up and down steep terrain and through other obstacles. The sawyers who carried the chainsaw, gas, oil and maintenance kit will cut down other trees and pesky plants like vine maple while the swampers make sure all the branches and shrubs are thrown in the black.

Sometimes, crews will be called on for initial attack (IA). Initial attack happens when a crew first responds to a fire and tries to stop it before it gets too large to control. When this is the case, a wildland firefighter's hours may be extended. At the very end of a twelve- or sixteen-hour shift, a spot fire may occur, and a crew may have to stay until it is out. I know this because it has happened to me a few times.

During an IA, hand crews will use their hand tools to pull fire apart and drown the fire with dirt. They also will have to dig out stump fires, root fires, grass fires and more. If you're lucky enough to have water, you will have to throw a house lay from the water source to the fire. These wildland firefighting techniques haven't changed in almost one hundred years.

What certifications does a person need to have in order to become a wildland firefighter (other than the free online FEMA training)? A person wanting to become a first-year wildland firefighter or basic firefighter II must take their S-130/190 basic firefighter training classes. These classes are taught by state and federal approved instructors who work for the U.S. Forest Service, Oregon Department of Forestry or privately owned training centers like the National Firefighter Training and Carding Association in Philomath, Oregon. The classes should cost less than $200 per person and can be taught by certified instructors with years of experience. This class should be much more than just a series of PowerPoint slides. In this class, a student will gain a basic knowledge of incident management organization, which is needed to understand the chain of command on any incident. A student should also learn how the different sizes of fire bases should work and how to attack a fire. Basic firefighting techniques are taught by

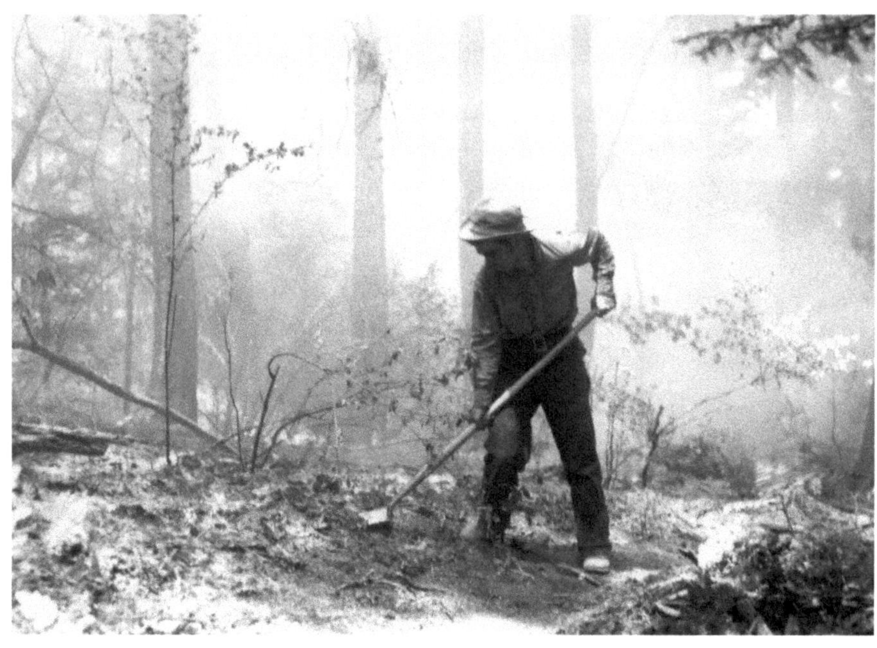

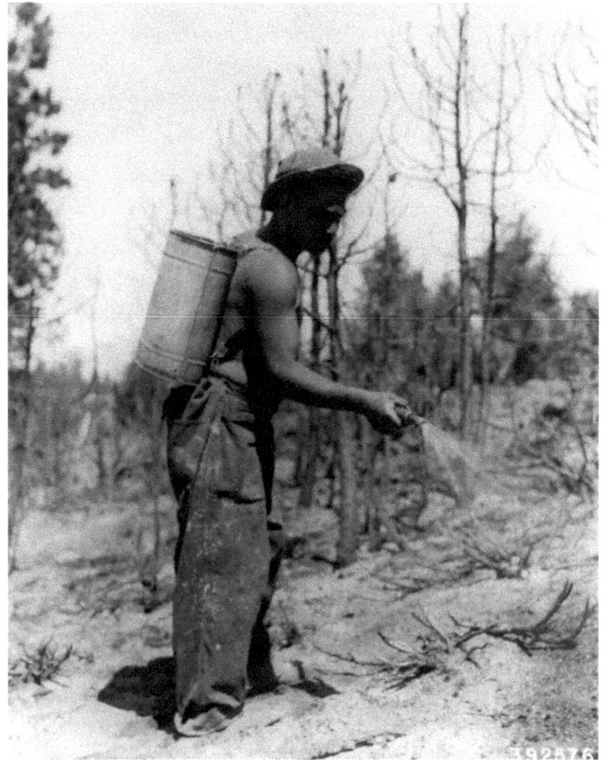

Above: A Civilian Conservation Corps young wildland firefighter works on dry mopping a log in the Smith Fire, 1938. *Courtesy of Forest History Center/Oregon Department of Forestry.*

Left: Backpack sprayer in action on the Mokst Butte Fire, 1936. *Courtesy of the Gerald W. Williams Collection at Oregon State University.*

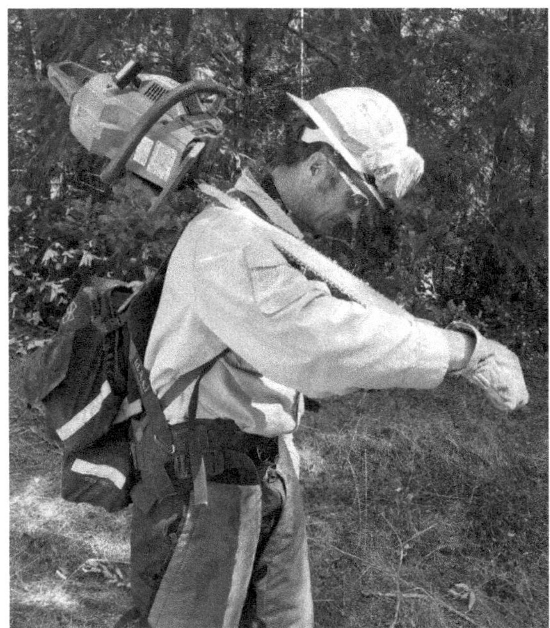

Engine boss and sawyer Eric Gatchell on the Gardner Complex, 2018. *Photograph by the author.*

spending hours out in the field digging fire lines and familiarizing potential firefighters with the wildland tools. This will include a drip torch (in case you have a controlled burn), diesel-powered pumps and different hose lays. Fire behavior, weather patterns and even leadership skills will be taught, but the most important topic is safety. During the time in the classroom, the instructor will go over fatal incidents on past fires and hopefully how to avoid them in the future. The safety lesson will be topped off by a timed exercise on the use of personal fire shelters.

Every wildland firefighter in the United States must wear an individual fire shelter, and every single one hopes he or she will never have to use it. A wildland firefighter only uses the fire shelter when his or her life depends on it. Fire shelters have been mandatory for all fires on federal land since 1977, and since then, they have "saved the lives of over 300 firefighters and prevented hundreds of serious injuries."[10] The outer layer of aluminum bonded to woven silica cloth reflects 95 percent of the radiant heat—heat directly from the flames. There is an inner layer of aluminum foil laminated to fiberglass that absorbs the convective heat, or heat from the hot gases and fuels that are actively on fire. These two layers are sewn together, and the air between the layers acts as insulation as well, but the shelter is no good against direct contact with fire. And even if there is no contact with flame, the glue of the shelter will burn away at around five hundred degrees Fahrenheit.

Familiarization with the individual fire shelter is mandatory, as are drills to see if the firefighter can deploy and cover him or herself fully with the shelter within fifteen to twenty seconds. It's very difficult for a firefighter to do this in training, but add the stress of a fire and unknown terrain and it can be near impossible.

Imagine discovering that you are surrounded by fire, and according to the fire behavior, weather and your squad boss, you will have to deploy your fire shelter. Your escape route to the safety zone is cut off, so now you will have to find some sort of clearing. Once you find a halfway suitable deployment zone, you will have to work furiously to scrape it down to mineral soil or burn all the fuel with fusees, because if you don't, it can combust inside your shelter. You push your muscles until they can't move any faster, and you have finally cleared a space; now, you throw your tools far away from you. The pack that you have been trained to keep so it will save your life, the hand tool that has become an extension of your body and every other lifesaving tool must be thrown far away, because at this point, they are nothing but fuel. You pull out the folded shelter, and the fire that's burning around you has created its own wind, so if you're not holding on tight, the shelter will kite and fly away from you. Finally, you shake it open, step inside and pull it over your head. The fire is a jet engine in your ears. You lie flat with your head away from the oncoming fire. Your boots and legs hold down the bottom of the shelter, your gloves hold the top and even though you are covered with leather and Nomex, the intense heat feels like it's burning away your skin. It feels like there's no air to breathe, so you do your best to dig a small hole in the dirt below your face with your gloved fingers. The heat is everything. The heat is your whole world, and you try to cut through it with memories of your loved ones and reasons why you shouldn't give up and die just yet. The heat is scorching and charring your skin at every point a piece of your body touches the thin aluminum that is separating you from hell. All you can do is wait, hope and pray it passes before the glue melts and the shelter comes apart.

It could be the fear, or maybe it's the pain, but some firefighters abandon their shelters during a burnover. During the Dude Fire in Arizona in 1990, wildland firefighter Curtis Springfield's shelter reached six hundred degrees and delaminated on one side, causing him to leave his shelter, according to a document written by the International Association for Wildland Fire for their 2009 Safety Summit. His fellow firefighters heard him scream, "I can't take it anymore," and he tried to escape the fire. While running, he hit another firefighter's shelter and caused that firefighter to sustain third-degree burns

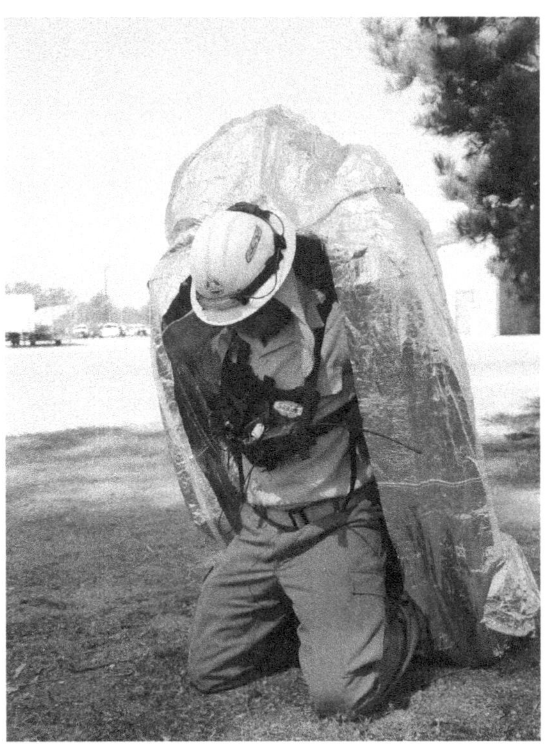

Right: Senior airman Ronald Skala, 30th Civil Engineer Squadron heavy equipment operator, demonstrates the use of a fire shelter on September 21, 2015, at Vandenberg Air Force Base in California. Constructed of aluminum foil, silica and fiberglass, a fire shelter is designed to protect the individual from smoke, heat and flames. *Photograph by senior airman Kyla Gifford, courtesy of the U.S. Air Force.*

Below: This is a crown fire on the Trout Meadow fire in 2007. *Courtesy of the U.S. Forest Service.*

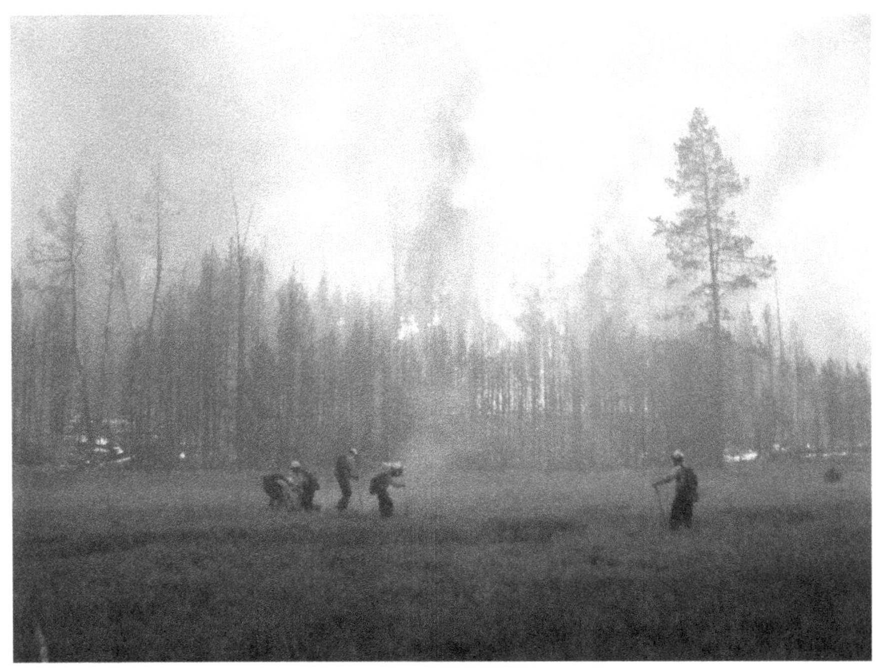

where his shelter touched his skin. Springfield only made it 150 feet down a dozer line before the hot air burned his lungs and he dropped and died.

New wildland firefighters should know the dangers of the job before they get on a crew. While deaths are rare, they do happen, and more fatalities happen today than ever before.[11] This doesn't necessarily mean that the job is getting more dangerous. It does mean that there are more wildland firefighters today than there've been in earlier decades. According to Ecowest.org, there have been over one thousand wildland firefighter deaths since the Big Burn of 1910. The National Interagency Fire Center (NIFC) reports that a total of forty-one wildland firefighters have died in Oregon since the NIFC started recording work-related fire deaths in 1910. These deaths can happen in a number of ways, including fighting fires, having a heart attack on the pack test, a vehicle crash or a training accident.

After classroom and hands-on training, a potential wildland firefighter must pass the pack test once each year. This test consists of a person of either sex carrying forty-five pounds for three miles within forty-five minutes, and the candidate must do this without running. Pack tests are administered by the people teaching the annual classes or the agency or company a person is applying to fight fire with, whether it be the USFS, the Oregon Department of Forestry or a private contractor. Whomever is putting on the test must tell the USFS where and when they are holding the pack test so a USFS representative has the opportunity to show up to ensure all standards were met during the test.

Once a person has completed the FEMA training, passed the required fire safety classes and come in under forty-five minutes on the pack test, that person is certified to fight fire for that season. Wildland firefighters call this having your "red card." The USFS, the Bureau of Land Management, the Oregon Department of Forestry, dozens of contractor crews and many more organizations look to hire able-bodied people with red cards in Oregon.

A new wildland firefighter, or FFT II trainee, shouldn't have any trouble finding an agency to work for today. The fire seasons this decade have grown longer and more violent. When searching for a crew to join, there are a few options. Most people will start out on a hand crew. A hand crew ideally consists of twenty wildland firefighters, and these firefighters are divided into squads of four or five. Each of these squads will have a squad boss. The squads make up a crew, and the crew has a crew boss. Many crews will have a designated sawyer and at least one swamper. Sawyers use chainsaws to clear areas, cut down snags (trees that can potentially fall on wildland firefighters) or fall fire-damaged trees.

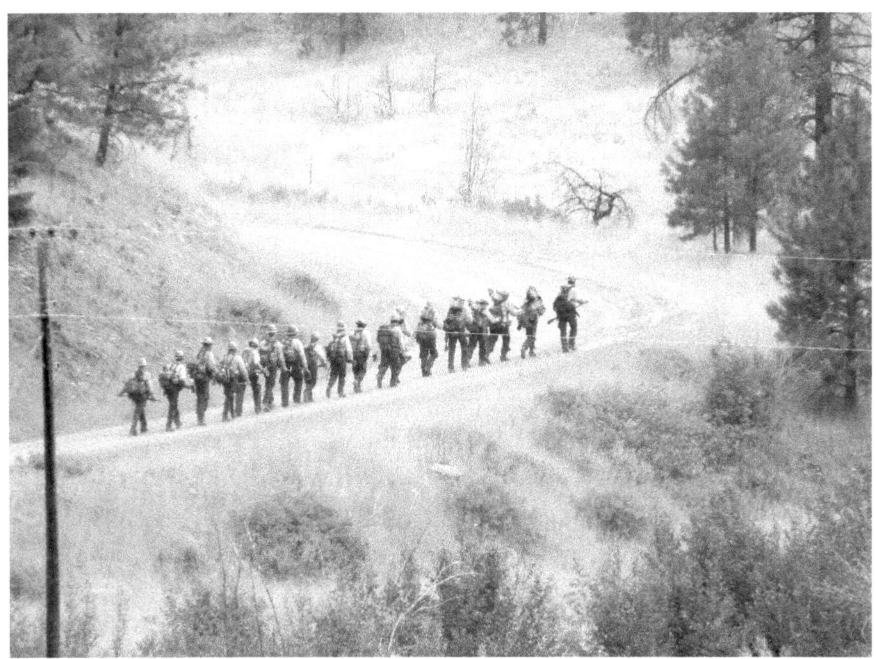

Line crew walking in the Willamette National Forest with their gear in 2017. *Courtesy of the U.S. Forest Service.*

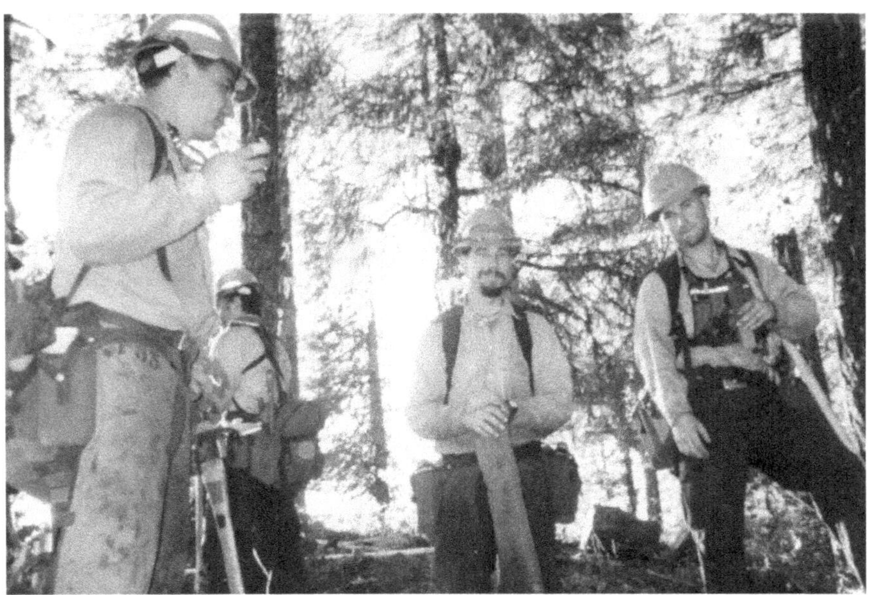

Zigzag Hotshot crew, 2014. *Courtesy of Tyee Burwell.*

As shown in the photograph below, Tyee Burwell was the sawyer for the Zigzag Interagency Hotshot crew on the Clark Fire near Lowell, Oregon. He needed to fall this tree because the fire had climbed the base and gotten into the branches. The burning of the tree created the potential hazard of branches falling off or the tree falling over onto wildland firefighters or future residents. Widow-makers and snags have always been some of the more likely and deadly hazards of wildland firefighting.

Every sawyer has at least one and sometimes up to four or five swampers depending on the assignment. Swampers clear the area of cut brush, pound wedges into trees to make the sawyer's job easier and keep an eye on any other hazards as the sawyer does his job. They also are taught how to be sawyers while on the job.

Once a person is a wildland firefighter for a few years and builds up knowledge of fire behavior, initial attack experience, pumps usage, saws, hand tools and much more, then that person has the opportunity to move to other types of crews if they want.

One of those types of crews drive on wildland fire engines. Engine crews consist of only three to four people depending on the size of the truck. These

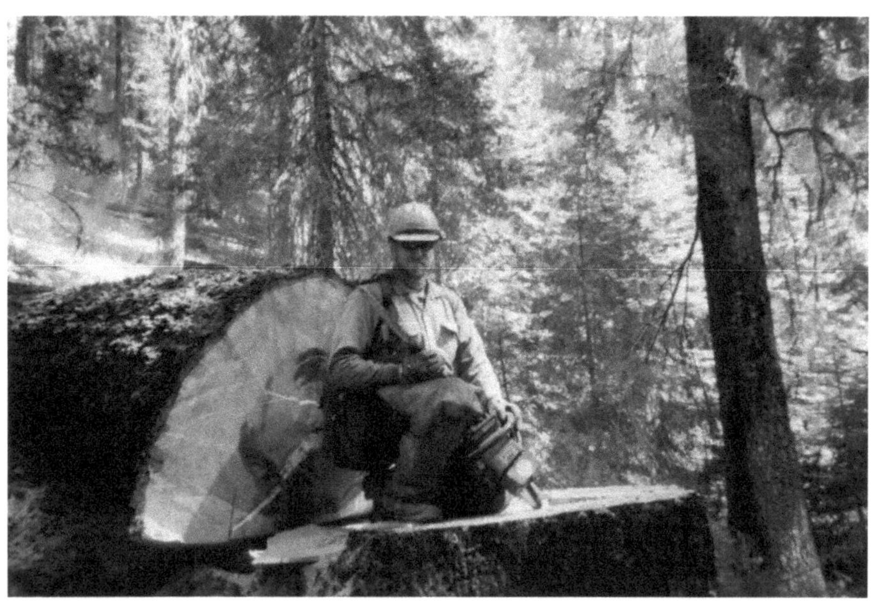

Tyee Burwell falling a large tree. Note how the lower right is ash. The tree was burnt, and it was determined to be a hazard, so the crew was told to fall the tree to ensure it did not become a safety issue. *Courtesy of Tyee Burwell.*

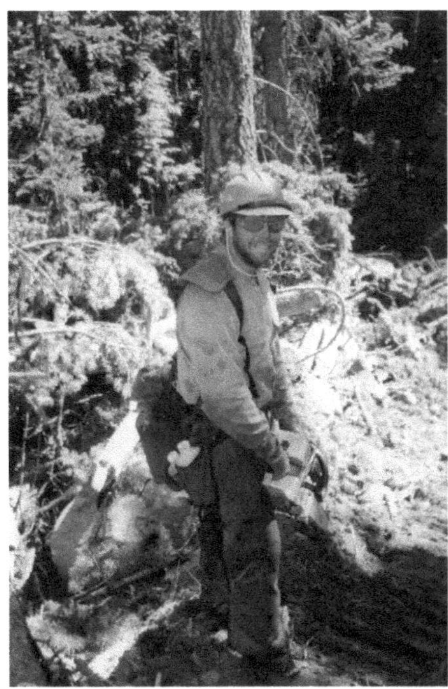

Left: Swamper on Zigzag Hotshots on Clark Creek Fire near Lowell, Oregon. *Courtesy of Tyee Burwell.*

Below: Civilian Conservation Corpsmen felling a snag on the fire line in the Siskiyou National Forest, 1936. *Photograph by Ernest Lindsay, courtesy of the Gerald W. Williams Collection at Oregon State University.*

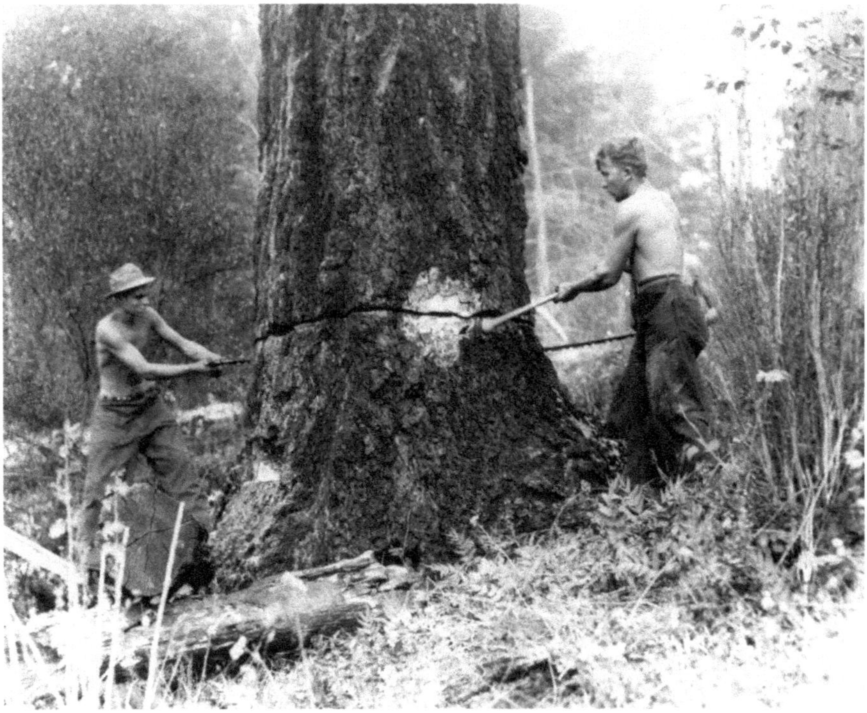

engines have large tanks of water, pumps and many different hoses on them, not to mention hand tools, saws and other essential equipment. There are different classes of engines, but none of them resemble the fire trucks you see fighting structure fires in an urban area. Water is harder to come by in the wildland setting. While there are many creeks, rivers and lakes to draft out of, water is still scarce compared to the number of fire plugs in a city or town setting.

Engine teams are a lot smaller than hand crews because you can only fit so many people and carry so much equipment in the trucks. Most of the engines fighting wildland fires are either Type 4, Type 5 or Type 6. A Type 4 engine has a minimum gallon capacity of 750 gallons. These are very large engines and can fit more personnel than the other types of engines, but what they gain in size they lose in maneuverability out on a fire. Not a lot of Type 4 engines are going to be climbing unimproved logging roads up to a fire in a remote area. Type 5 engines aren't as large and have tanks that can hold anywhere from 400 to 750 gallons, and they usually have a crew of four people. Type 6 engines have smaller tanks (150 to 400 gallons), but

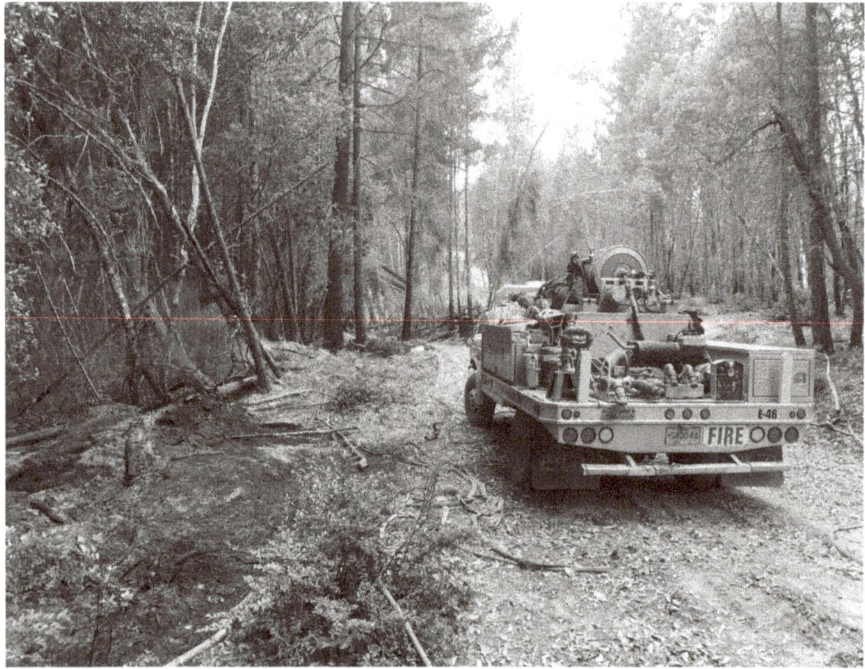

Type 6 engine from West Coast Wildland Strike Teams patrolling the black and putting out smokers on the line at the Happy Dog Fire in Southern Oregon, 2017. *Photograph by the author.*

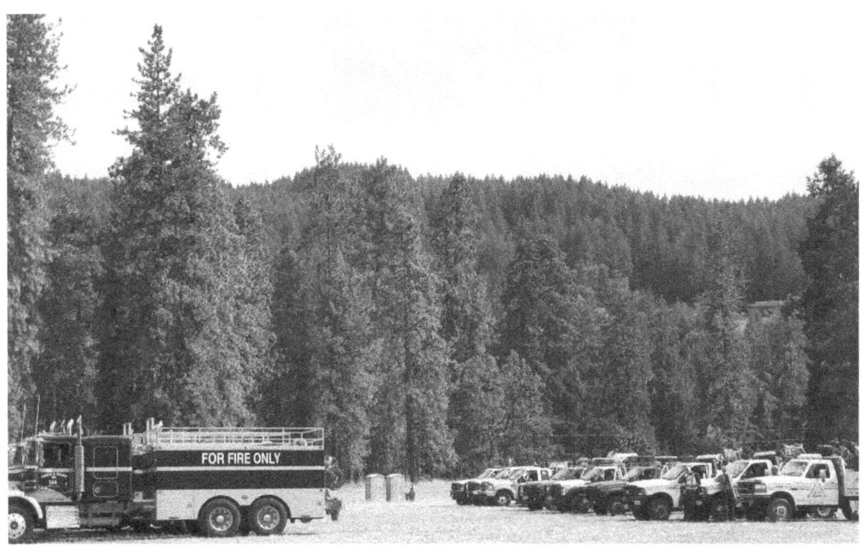

Type 5 and Type 6 engines lined up at the fire base on the Flounce Complex in Southern Oregon, 2017. *Photograph by the author.*

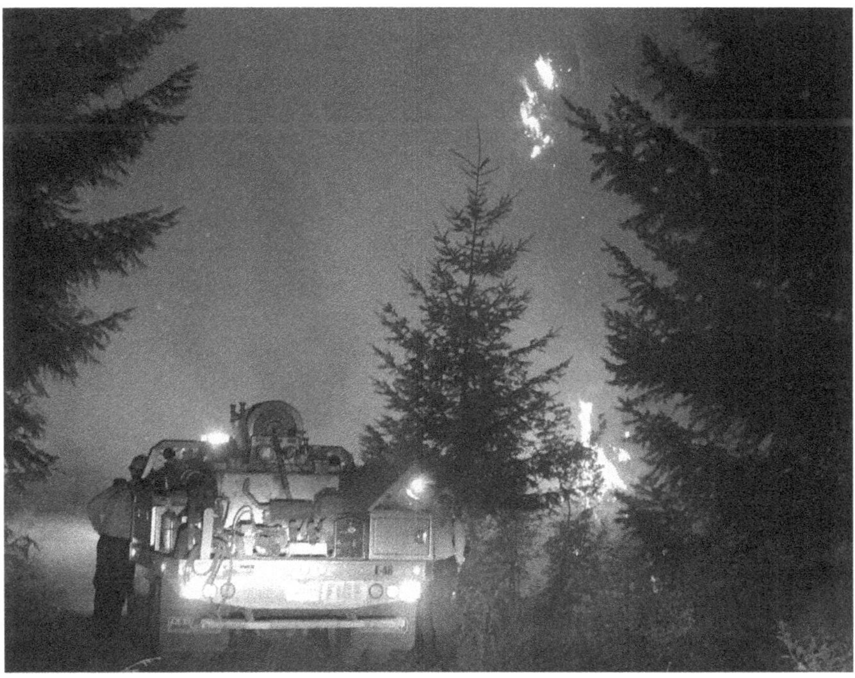

Type 6 engine about three miles up an unimproved logging trail at the Happy Dog fire in Southern Oregon, 2017. The crew is getting prepared to knock the fire out of the tops of the trees and off the road. *Photograph by the author.*

they make up for their lack of water with an increased ability to maneuver on small logging roads and dozer trails. Type 6 engines usually have three or four people on them. Type 6 engines can get to places a person wouldn't believe. I know this from personal experience of driving one on the Flounce Fire in Southern Oregon in 2017.

Experienced wildland firefighters also have the option to join a helitack crew. Helitack crews usually work for the USFS or the Oregon Department of Forestry. These are teams of wildland firefighters who are flown to the fires by helicopter. Helitack crews are used when speed is vital. For example, if a lightning strike is reported during the late spring or summer under extreme fire danger, a helitack crew may be called to respond. They are flown out to the site, and they will land if possible; if not, the team rappels down ropes. After they land, any equipment they will need to fight the fire is lowered or dropped down to them. Then, the team will load up and hike to the fire. Once there, they will make the decision to attack the fire by dry mopping, falling trees, firing operations or digging a fire line around it. If the fire is too big, they will then size up the fire and call for appropriate resources.

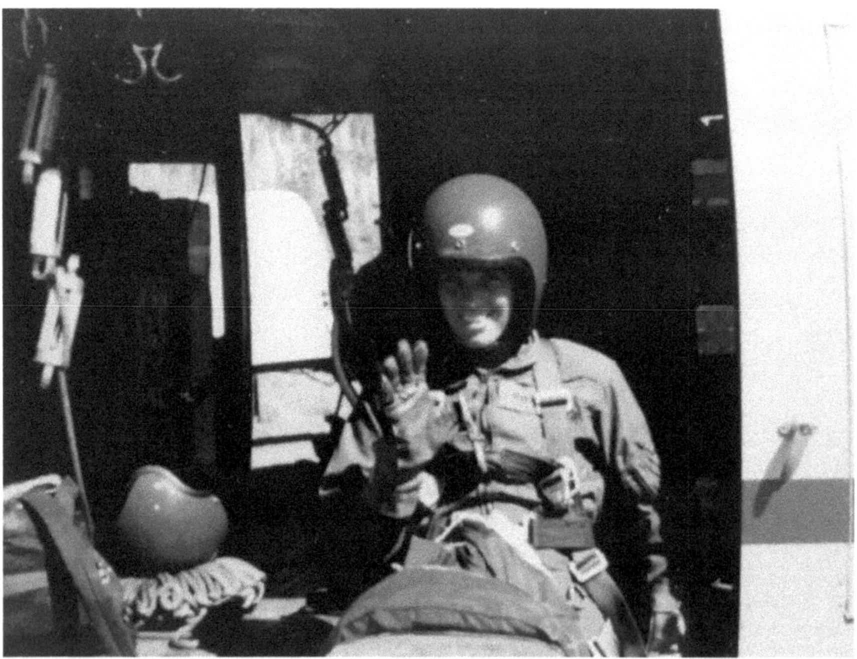

Rappeller Maria Luisa Ruiz, one of the first female helitack fire rappelling firefighters. *Courtesy of Valerie Rapp.*

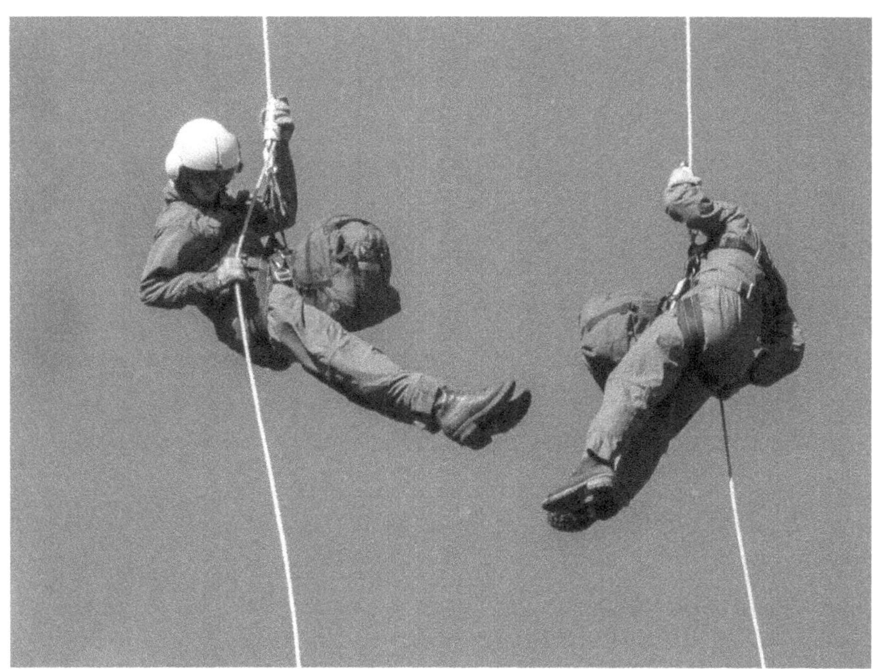

Veteran wildland firefighter rappellers use various techniques to negotiate the wind from the helicopter rotor blades above them and away from tree branches beside and below. This photograph is from May 12, 2014. Rappel crews may be utilized for large fire support, all hazard incident operations and resource management objectives. *Photograph by Lance Cheung, courtesy of USDA.*

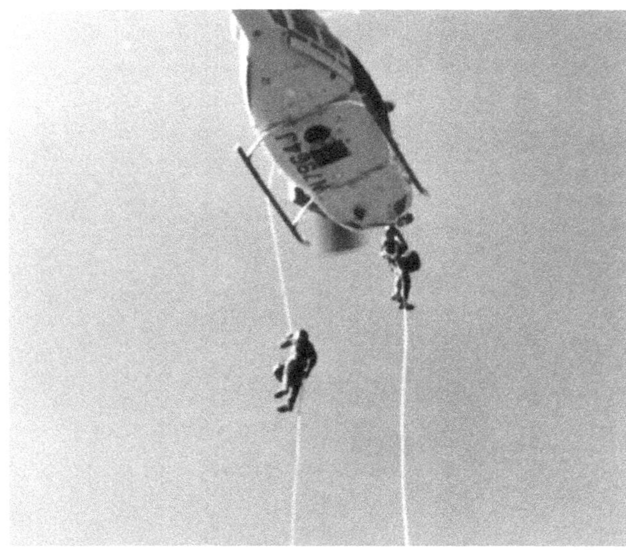

A 250-foot proficiency rappel held in the late 1970s by rappel base at Oakridge Airport in Oakridge, Oregon. *Courtesy of Valerie Rapp.*

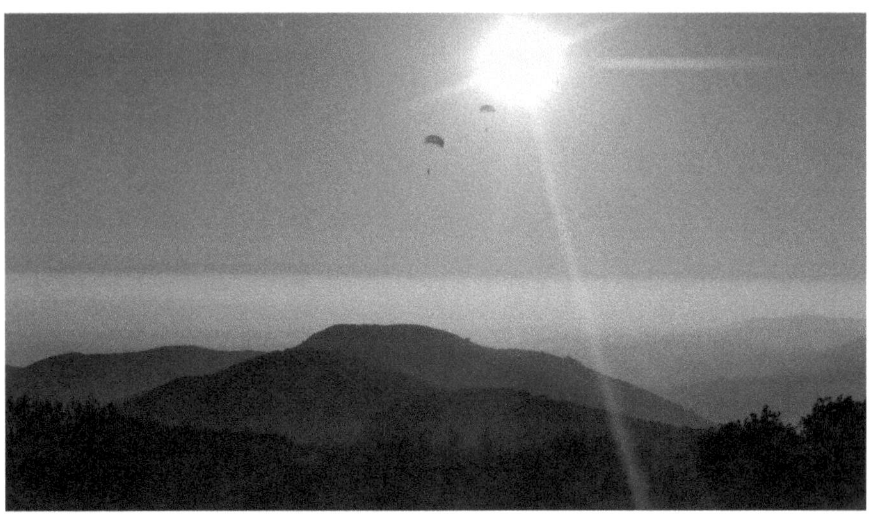

Smokejumpers jump with the equipment needed for an initial attack on a fire. After they do what they can, they must carry all their equipment out with them. Here are two descending to fight a fire in 2016. *Courtesy of Inciweb.*

Smokejumpers are another tool in fighting wildland fire, and this job is just what it sounds like. Smoke is sighted in a forested area, and these firefighters load into a plane, fly to the area and jump. The smokejumping program started in 1939 in the Pacific Northwest. The first jump was in 1940 into the Nez Perce National Forest in Idaho.[12] Today, smokejumpers are a national resource, and they travel all over the country. There are approximately 270 smokejumpers who operate out of eight smokejumper bases across the United States, including one in Redmond, Oregon.

The Redmond Smokejumper Base officially opened in June 1954, and later that same month, eight men parachuted into the Dixon Fire on the Warm Springs Indian Reservation. The airfield on which the base is located was once a training airfield for pilots in World War II. The Redmond Smokejumper Base averaged 350 jumps per year into about eighty-six fires per year for the last ten years.[13]

When trying to help people outside of incident command understand the different types of crews on a fire, the USFS describes hand crews as the infantry of firefighting. Helitack and smokejumpers are the air assault and airborne divisions. So what do they consider the special forces of modern firefighting? Well, if you want the most elite crew of incredibly hardworking firefighters, you call in the interagency hotshot crews (IHCs).

Hotshot crews are filled with the best wildland firefighters, many of whom have years of experience before being accepted into an IHC. In addition to the annual retraining and pack test every firefighter goes through, hotshot crews set individual physical standards that include running a mile and a half in at least eleven minutes, doing forty sit-ups in sixty seconds and twenty-five push-ups in sixty seconds, chin-ups and more. These are the minimum standards set by the USFS, but individual crews can set higher standards, and most of the time they do. IHC members also get a lot more classroom time and cover subjects such as characteristics of extreme fire behavior, hazard tree safety, leadership development and other classes about how to work with fixed-wing aircraft and helicopters, to name just a few.

Hand crews, engine crews, helitack crews, smokejumpers and hotshot crews are just some of the options for those looking for a career in wildland firefighting. Firefighters can also stay on track to become a crew boss, division supervisor or even incident commander, but they have to get through the season and have someone sign off on their task books first. After a wildland firefighter has been certified, passed the pack test and been hired, they are given a task book. This is a book of tasks a wildland firefighter must complete over at least three fires and fifteen days of working on a fire. Most of the time, it takes a person a lot longer than that to fill out a task book, but that is the minimum. Once a wildland firefighter's FFT II task book is filled out, the firefighter is given his or her FFT I task book to fill out, then he or she can move on to other positions according to his or

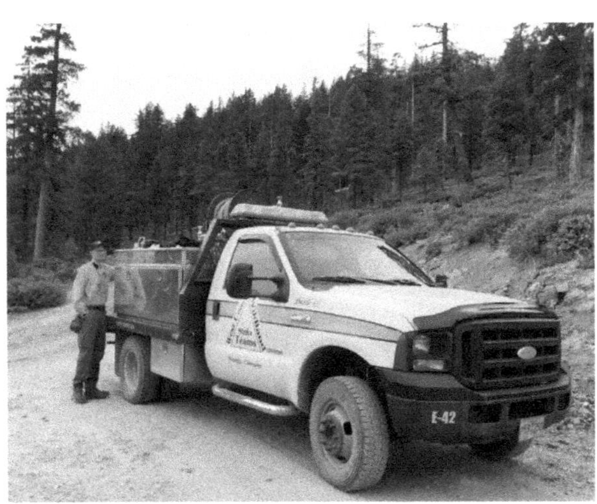

Wildland firefighter Marques Gatchell on a Type 6 Engine for a contract crew. Gatchell is one of many second-generation firefighters who fights alongside a parent. His father is an engine boss on a different Type 6 engine. The job is dangerous but rewarding enough that many firefighters bring on their children. This photograph was taken on the Garner Complex, 2018. *Photograph by the author.*

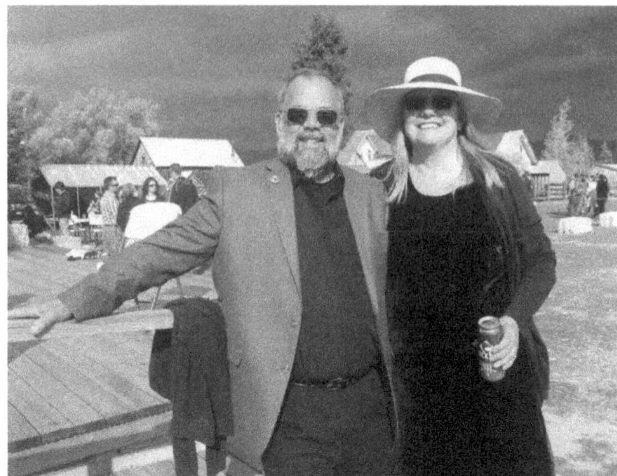

Left: Sam Swetland and his wife, Cookie, at home. Sam has fought fires for forty-two seasons. *Courtesy of Sam Swetland.*

Below: Tyee Burwell falling a large snag, 2003. *Courtesy of Tyee Burwell.*

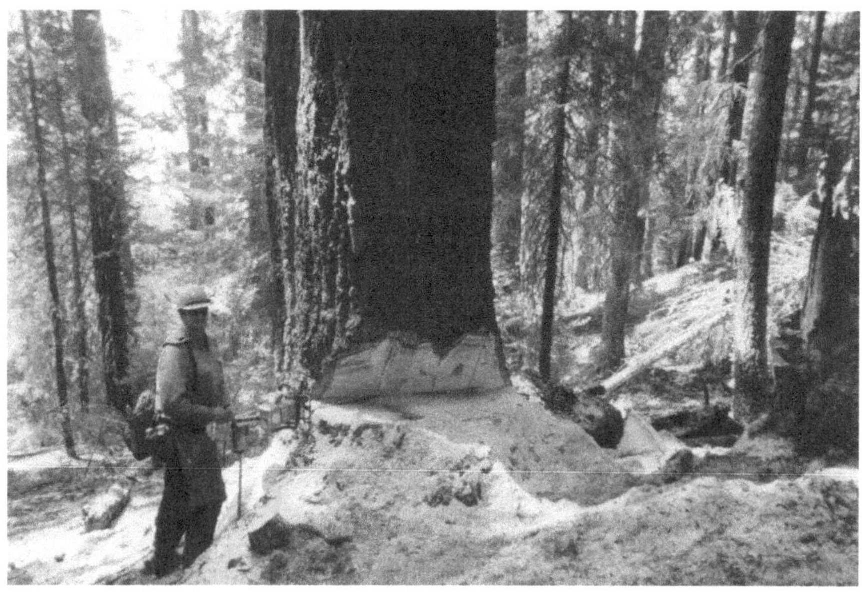

her specific interests. A FFT II can grow throughout his or her career and supervise all air resources like Sam Swetland, become an engine boss like Keith Hergert or become a crew boss on a hotshot team like Tyee Burwell. Bigger fires have dozens of leadership positions to fill, so it really depends on what the wildland firefighter finds interesting.

Chapter 5

THE CAUSES AND SIZES OF WILDLAND FIRES

The great state of Oregon has more trees than any other state except Alaska. There are small states on the East Coast that have a larger percentage of trees compared to the size of their state, but as for the number of trees, Oregon beats them all. The forest, wilderness and trees are a part of Oregon culture. People from all over the world come to see the beauty of the state. Outdoor recreation is a way of life for those who are lucky enough to live here, but it can be dangerous. According to the National Park Service, 90 percent of the wildland fires in the United States are caused by humans.[14] The government has recorded 1.5 million wildfires between 1992 and 2012, and according to its records, 25 percent of these were caused by people burning trash debris and 22 percent were started by the use of power equipment; heat from the exhaust of ATVs, motorcycles and other recreational vehicles; smokers; campfires and other recreational activities.[15]

Many fires have been started by logging or other work in the woods, and sparks coming off a railroad track also cause fires. Dozens if not hundreds of fires are caused by lightning strikes every year. There are too many more sources to list them all, but even the smallest spark can become the largest fire depending on weather and fuel conditions.

A small spark can become a fire that rages for hundreds of thousands of acres. Every time a fire is reported, a crew is deployed to size it up and call for the resources that will be needed to put it out or contain it. Each different type of fire requires a different type of leadership. The fires range

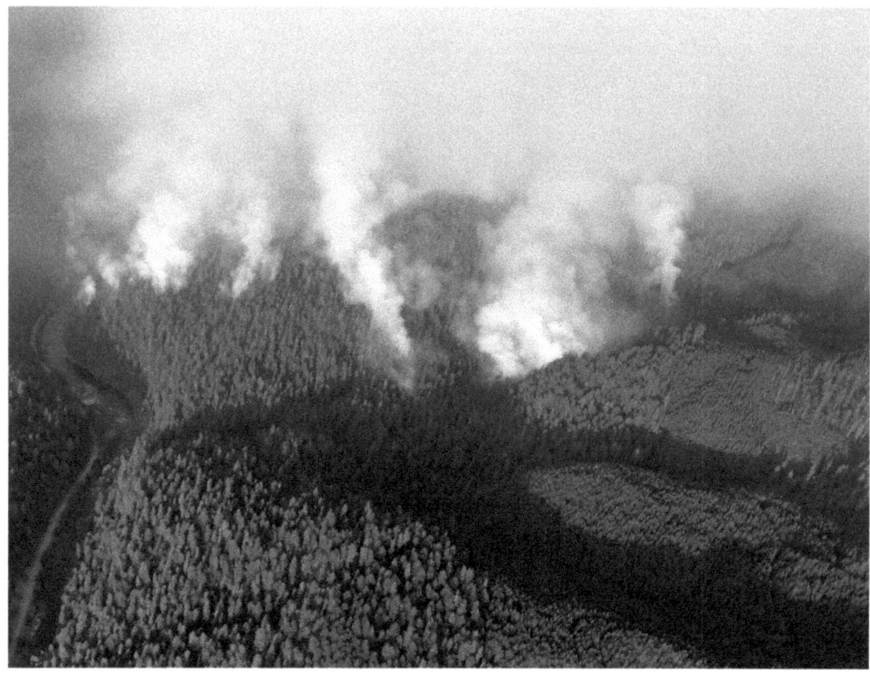

Smoke column from North Umpqua fire in 2017. *Courtesy of the Bureau of Land Management.*

from Type 5 to Type 1 according to the Incident Command System. A Type 5 Incident is just a local resource and will require anywhere from two to six firefighters. This could be related to an unmanned campfire or have a crew chasing lightning strikes in the woods. This kind of fire is usually put out by first responders within a matter of hours after arriving on the scene.

A Type 4 Incident will also require only local resources, but no one activates the command staff and general staff functions needed for larger fires. There may be a briefing for incoming resources but not a real daily incident action plan (IAP) like those that will be distributed on bigger fires. Of course, Type 4 and 5 Incidents can grow to Type 1 fires, but until they reach a certain size and require a certain amount of wildland firefighters, they will be fought by the resources available in the immediate area.

When a fire grows and needs a larger number of wildland firefighters, support services are going to be needed. This hasn't changed over the years. Fighting fire is a very physical activity, and food and water are very important. Bigger fire camps will have mobile kitchens and packed lunches for the wildland firefighters.

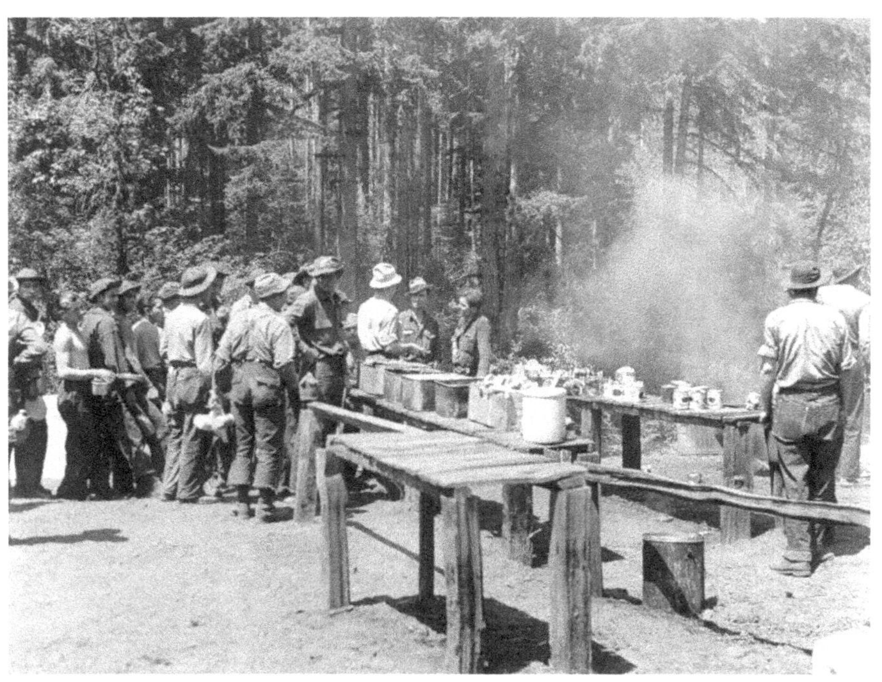

Chow line at the Smith River Fire, 1938. *Courtesy of Forest History Center/Oregon Department of Forestry.*

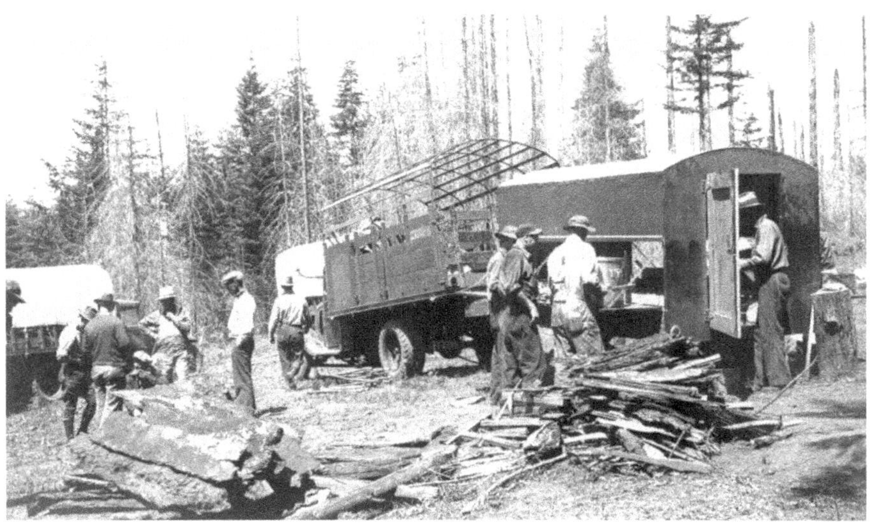

The kitchen trailer in this photo was built by the Civilian Conservation Corps at Camp Nehalem. Here, it is being used at the 1939 Scappoose Fire. *Courtesy of Forest History Center/Oregon Department of Forestry.*

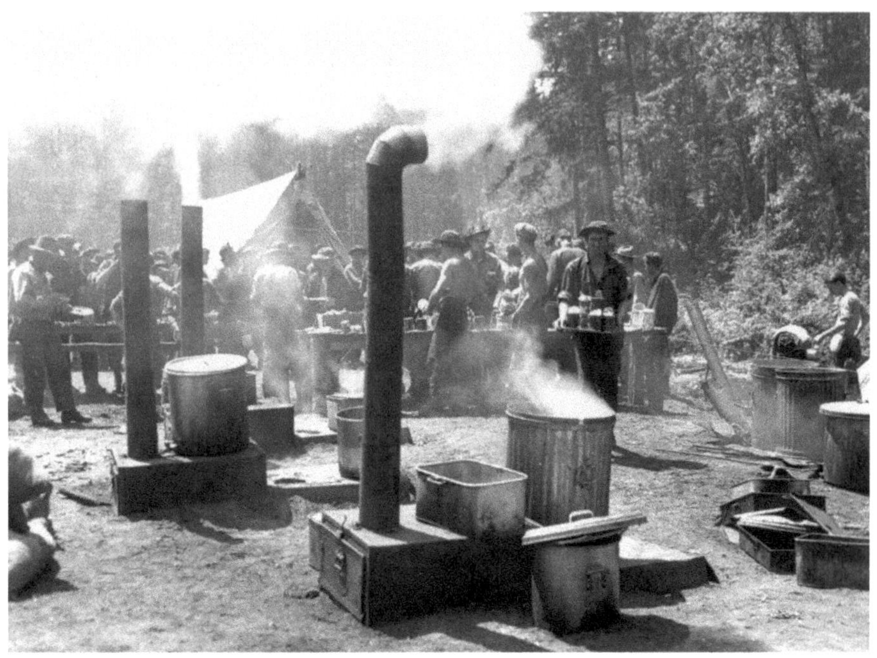

The "kitchen" cooking up some great food to fill the wildland firefighters before they battle the Smith River Fire, 1938. *Courtesy of Forest History Center/Oregon Department of Forestry.*

Mobile kitchens have been a part of a fire base for a long time, but whenever a fire base offers breakfast and dinner, they usually issue bag lunches. Before every shift, someone on a crew (usually the FFT IIs) will have the responsibility to check out lunches, bottled water and sports drinks from a designated area on the fire base. The firefighters will carry the lunches and sleeves of water and sports drinks back to their crews to divide them up to be carried throughout the day until lunch. Being on the fire line and actively putting out the fire is very important, but so are breaks. The firefighters have to find time to eat and rest when possible.

When a fire is taking up most if not all the local resources, and a command staff is required, the fire has moved to a Type 3 Incident. This fire will have an incident commander, one or more division/group supervisors and a significant amount of wildland firefighters. There will be formal morning and evening briefings, and resources will be divided into several task forces/strike teams. This means that there will be hand crews, engine crews and maybe some heavy equipment or hotshots. Every morning at the briefings, the IAPs will be handed out to all leadership

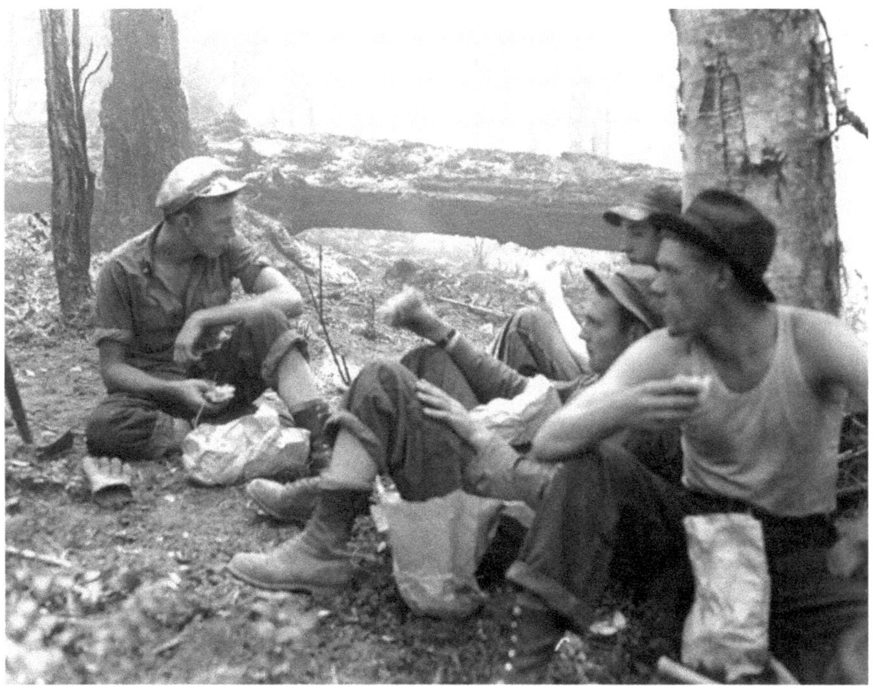

Civilian Conservation Corps crew taking a break on the Smith River Fire, 1938. *Courtesy of Forest History Center/Oregon Department of Forestry.*

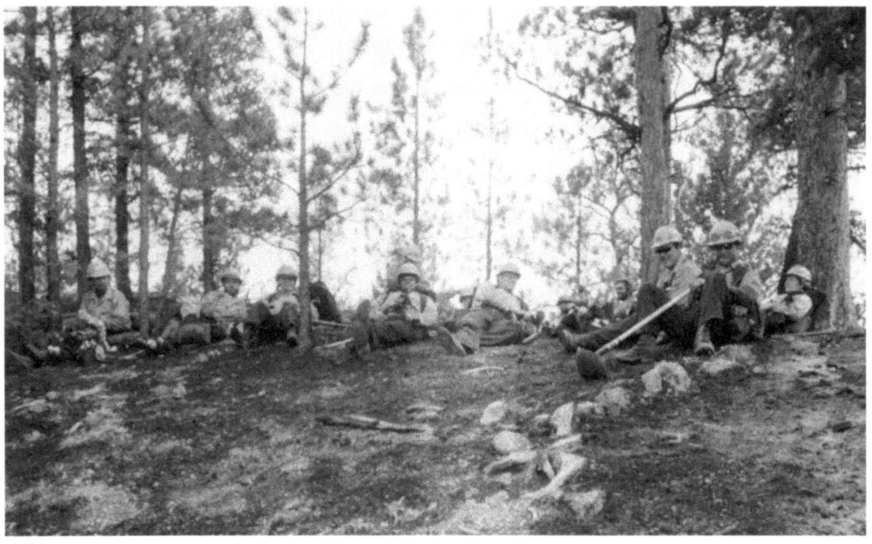

Zigzag Interagency Hotshot crew taking a break on a fire, 2007. *Courtesy of Tyee Burwell.*

in order to let everyone know about the main plan of attack and also make sure everyone knows what the other groups are doing. At this point, a local park, school or campground will be designated as the fire base. The wildland firefighters put up their tents, and it may look similar to a campground at a multiple-day music festival.

Type 2 Incident teams plan on being on a fire for a while. A normal fire deployment for a wildland firefighter lasts two weeks. During that time, there are no days off, and every firefighter works twelve- to sixteen-hour days. Some crews work the day shift and some work the night shift, and on some fires, there will be an additional swing shift. After working for two weeks, the crew has the choice of taking off one day and going back to work for seven more days, or the crew can take off two days and going back to work another fourteen-day stint. This is true for firefighters ranging from a hand crew consisting mostly of FFT IIs all the way up to the leadership—even an incident commander must rotate off the fire. Many times, the whole leadership team rotates off the fire at the same time. The incident base and all the wildland firefighters will continue to

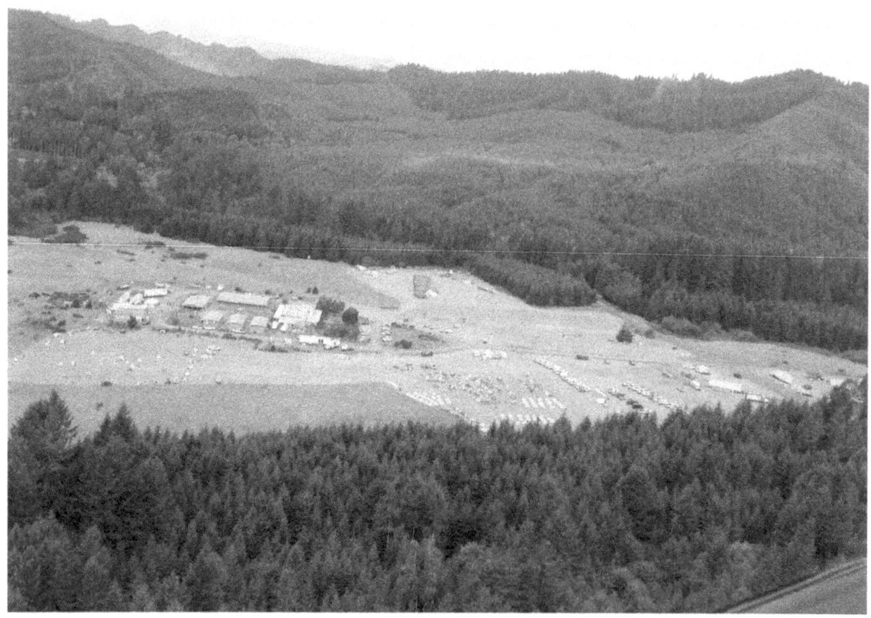

Type I Incident Fire Base for the Yellow Point Fire, twenty-five miles west of Cottage Grove, 2015. *Courtesy of the U.S. Forest Service.*

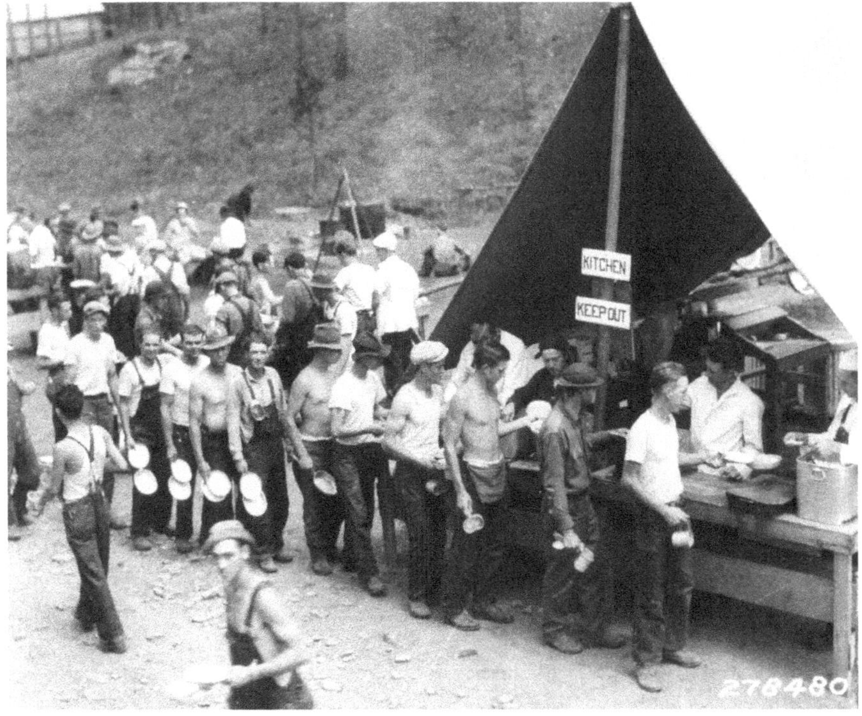

Fire camp kitchen at a CCC camp, 1933. *Courtesy of the U.S. Forest Service.*

function as long as they're needed. They will either put out the fire and grid the black, contain the fire or, at the very least, control the direction of burn and let it go into the wilderness until the weather puts it out.

The largest incidents are Type 1. On a Type 1 Incident, the fire is divided into multiple divisions, often with more than five hundred operations personnel working on the fire and personnel totals over one thousand. There will be aviation operations, heavy machinery, hotshot crews, engines of different sizes, laundry services, mobile food services, equipment exchange services and more moving parts than most people can count. These are the large complex fires that may not be contained until the seasons change and it starts raining or snowing.

On large fires, the incident team may work with the community and use a school, fire department, park or some other area as their headquarters. A fire base must be so far away from the fire—or fires, in the case of a complex—that there is very little or no worry that they will have to move. Some larger fires will call for thousands of wildland firefighters and all the

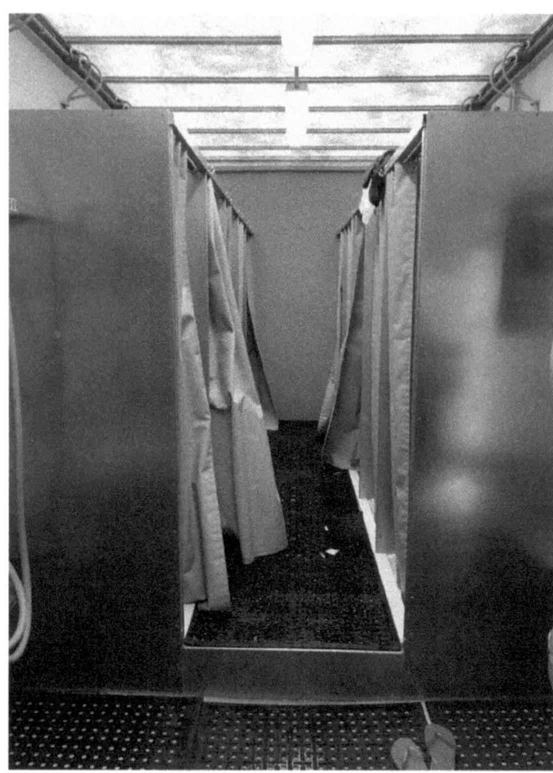

The inside of a shower tent on a Type I complex fire. *Photograph by the author.*

equipment needed to support them. Some Type 1 fires have had multiple food trucks, shower trucks, supply points, et cetera.

Every morning and every evening, before the wildland firefighters and other resources are sent out to fight the fire, they must eat, inspect their equipment, take care of personal hygiene and go to briefings to find out what their assignment is for that shift.

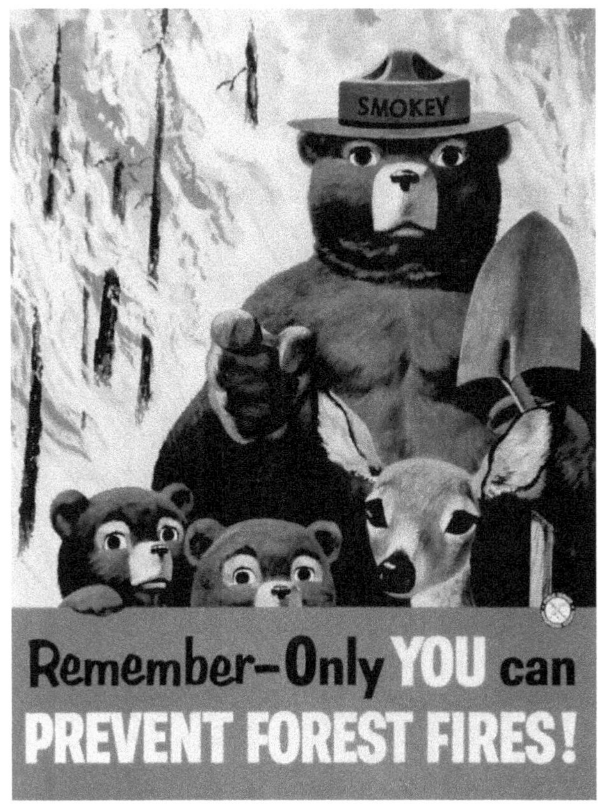

Left: Smokey Bear poster from 1957. *Courtesy of the U.S. Forest Service.*

Below: A crowning fire raging. This is a good example of a crowning fire. This type of fire is very intense and throws off heat that can be uncomfortable hundreds of yards away. *Courtesy of the U.S. Forest Service.*

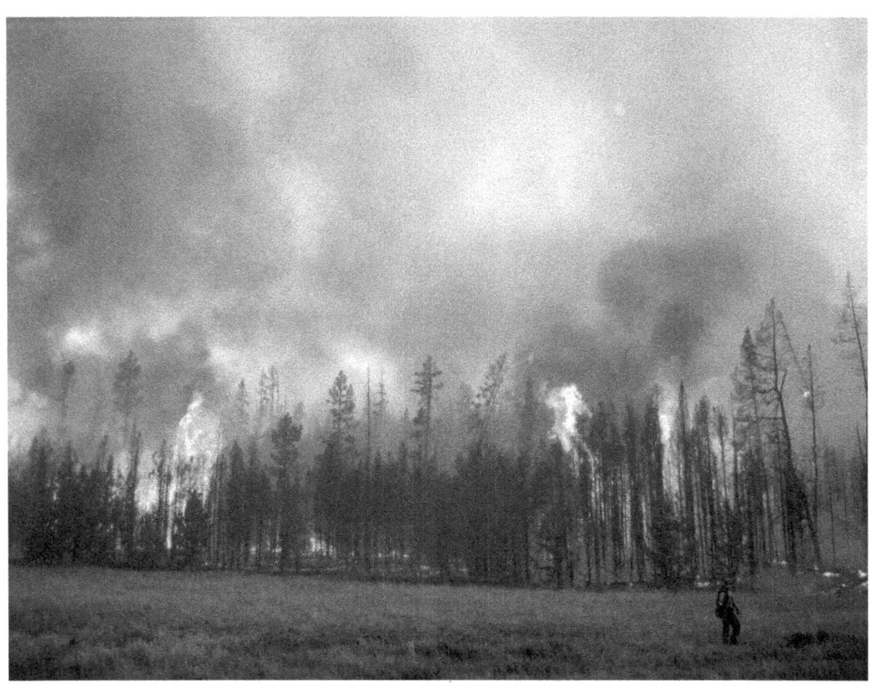

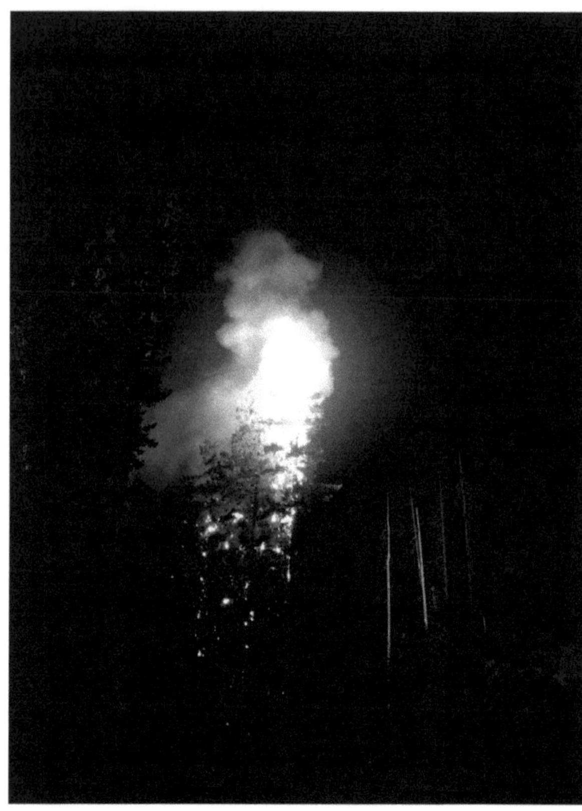

Left: The fire from a controlled burn climbed these one-hundred-foot firs. The fire ignited the "old man's beard" (*Usnea barbata*) that grew on the trees. This is a very combustible lichen, and once the fire climbed to the top, it burned and boiled the sap, causing the tops of the trees to explode. This caused spotting fires even though there was no wind. *Photograph by the author.*

Below: Keith Hergert on night shift on the Taylor Fire in Southern Oregon, 2018. *Photograph by the author.*

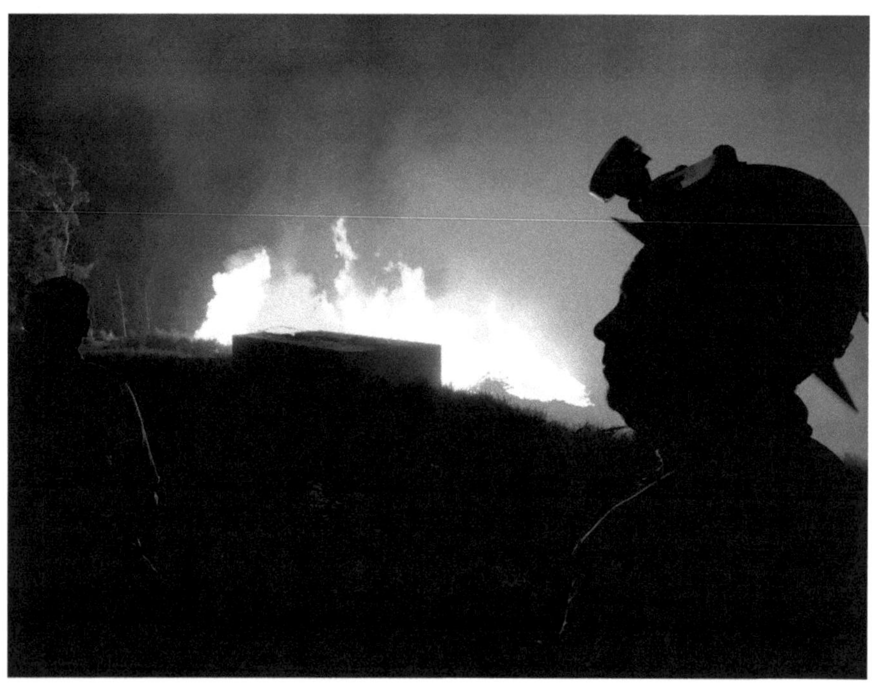

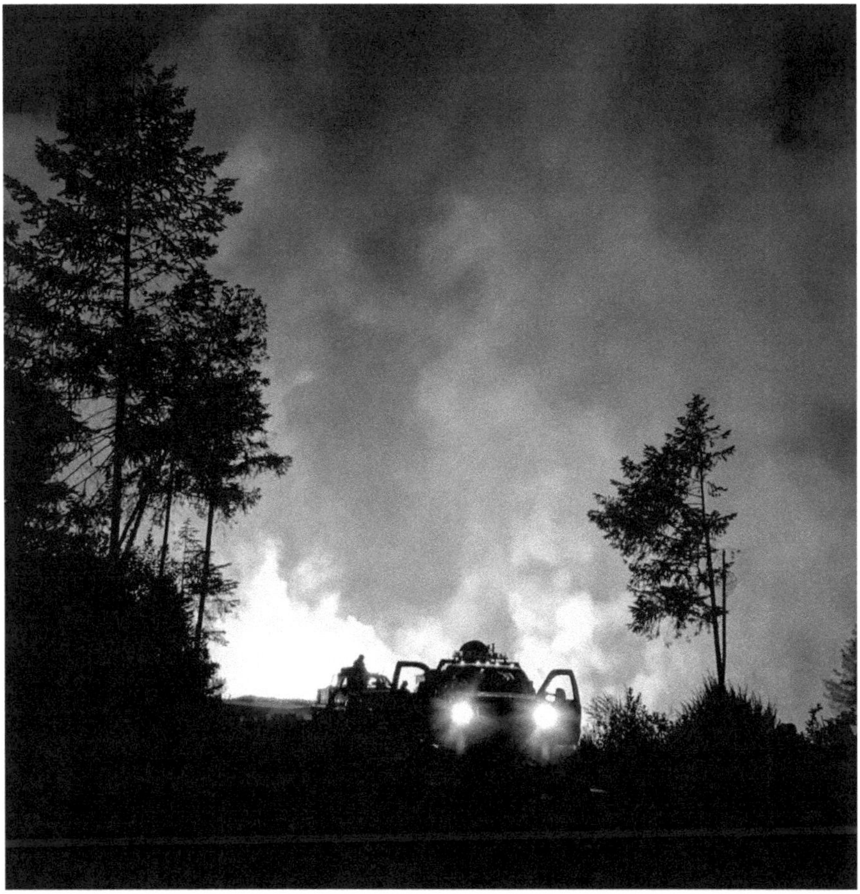

Above: A Type 6 engine fights a wall of flame on the Taylor Creek Fire in Southern Oregon, 2018. The fire started with a lightning strike on July 15 and burned over 52,000 acres until it was contained on October 21. Although the fire burned in a residential area with multiple recreational parks, firefighters were able to keep property destruction at a minimum. *Photograph by the author.*

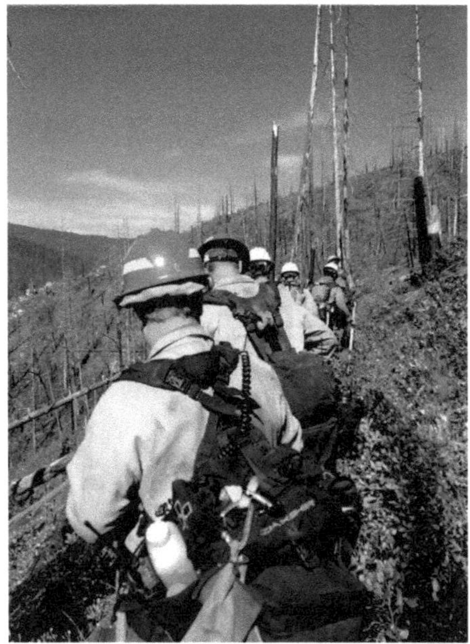

Right: U.S. Forest Service crew walking to a wildfire in 2015. Photograph by U.S. Forest Service strike team leader Ryan Otto. *Courtesy of Ryan Otto.*

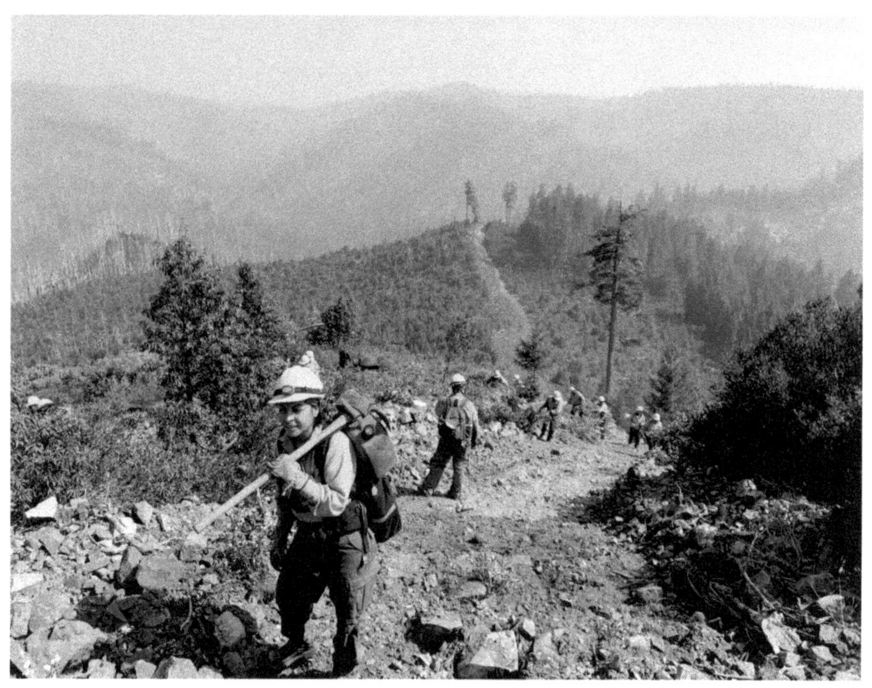

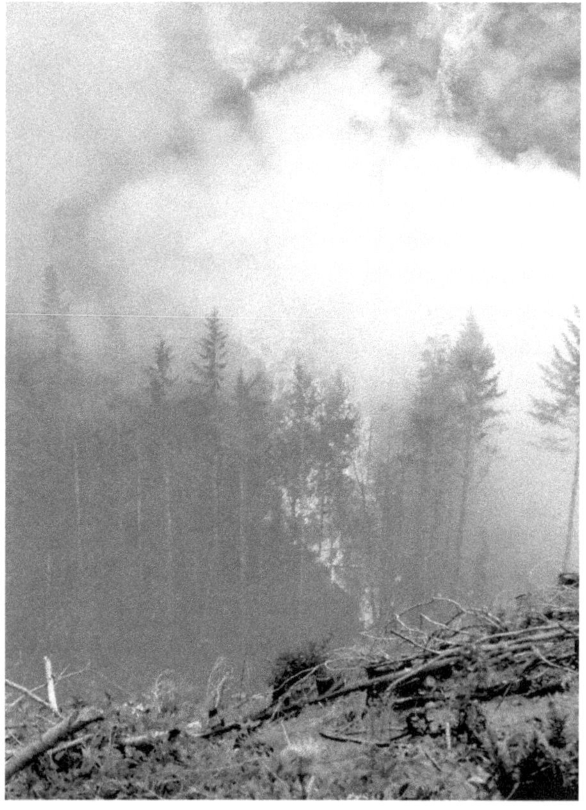

Above: Hand crews brushing fire line along Chrome Ridge east of Chetco Bar Fire, 2017. *Courtesy of Inciweb.*

Left: Yellow Point Complex, 2014, during a crown fire, in which the flames climb to the top of the trees and burn with a furious intensity. *Courtesy of the U.S. Forest Service.*

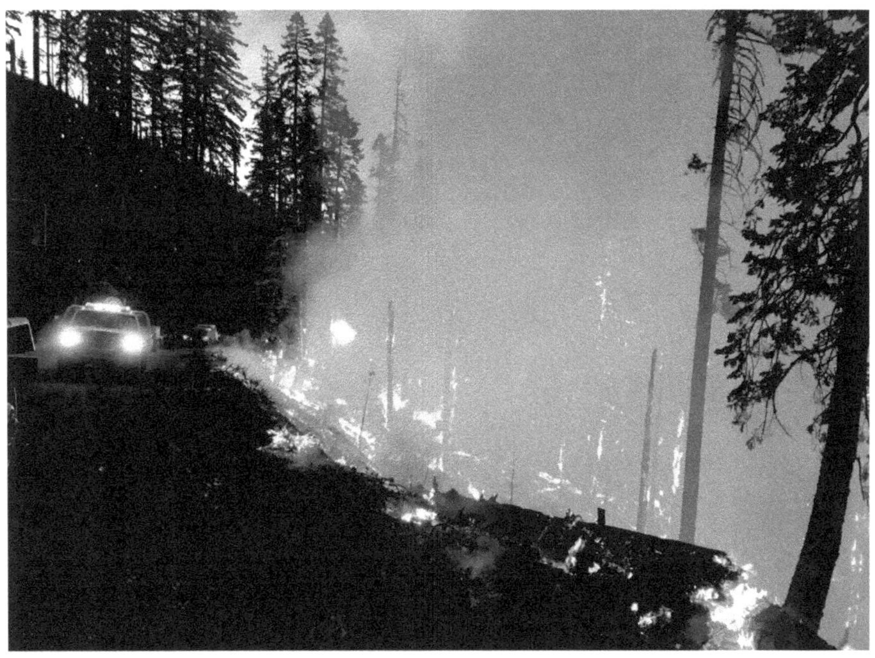

Active fire behavior with the fire burning up a ridge to the road on the Klondike Fire, 2018. *Photograph by the author.*

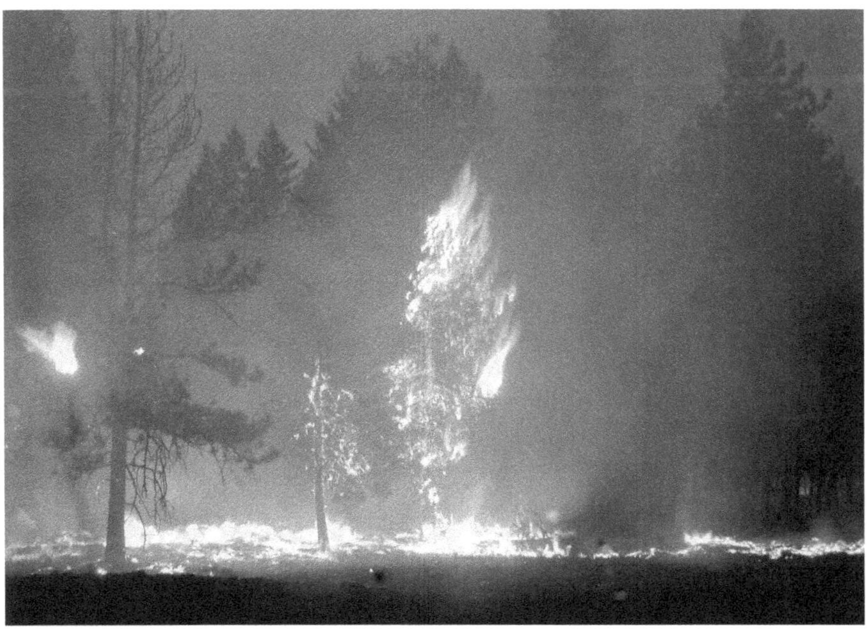

Here is a good example of a variable canopy. This is one of the dangers of wildland firefighting. A grass fire can easily heat larger vegetation and turn into a brush fire. Brush fires heat and climb into smaller trees and eventually climb larger trees. Once a fire fully engulfs entire trees, there is very little a wildland firefighter can do other than evacuate the area. *Courtesy of the U.S. Forest Service.*

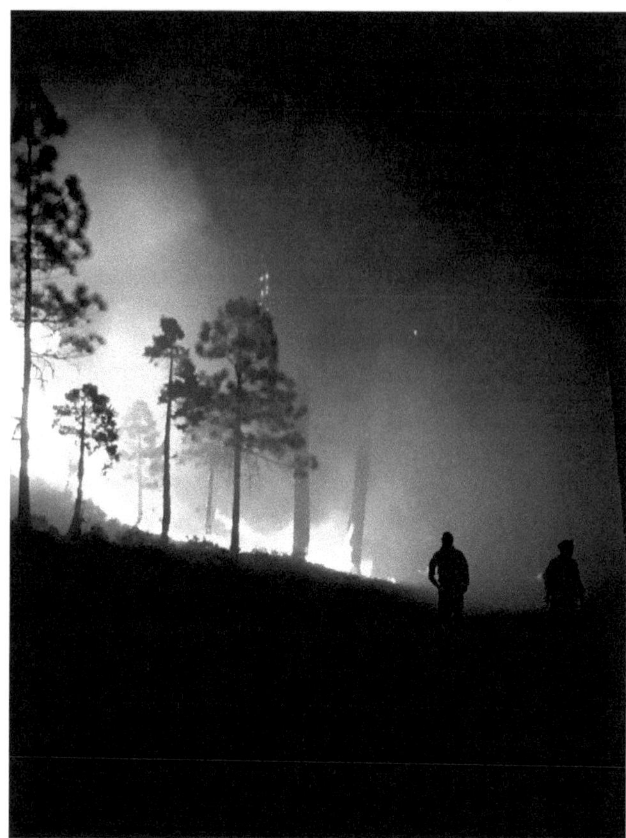

A photograph of the author (*left*) walking the fire line on the Klondike Fire with engine boss Keith Hergert in 2018. This fire was started by a lightning strike on July 15, like many others in Southern Oregon that day, but this one burned 175,258 acres and became the largest in Oregon that season. This fire burned into November. *Courtesy of Stephanie Kuykendall.*

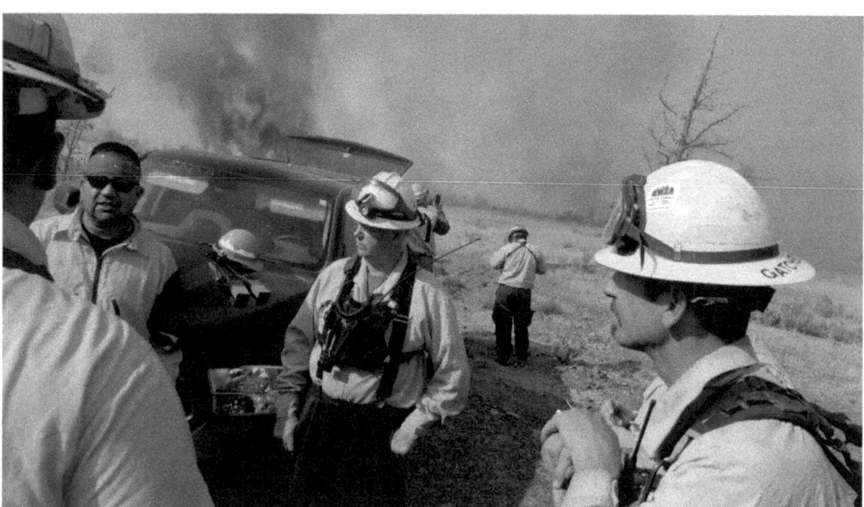

Wildland firefighters at a safety briefing at Warms Springs, 2016. *Photograph by the author.*

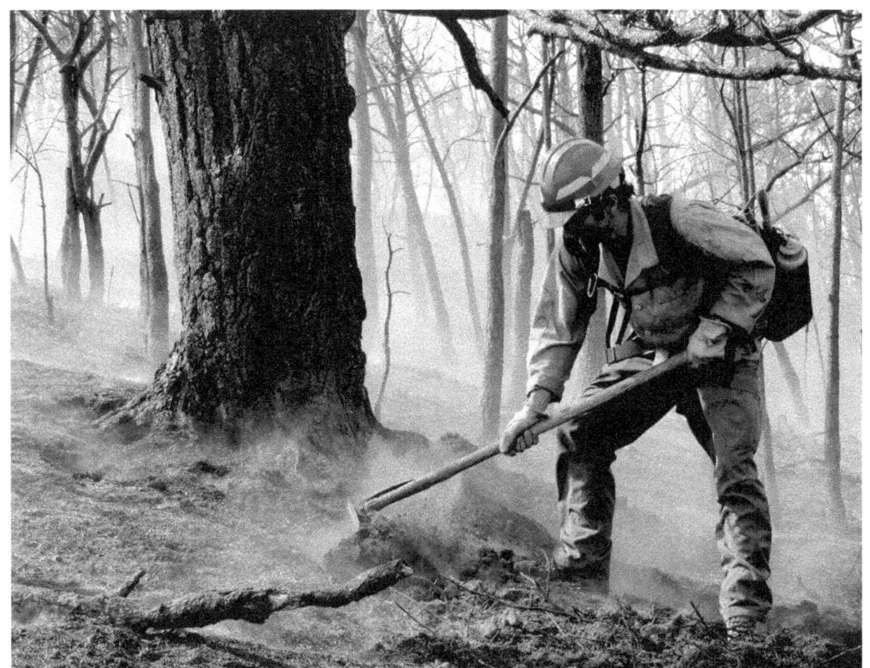

Above: A wildland firefighter digging out a root fire at the Stouts Fire, 2015. When the fire has burned over an area, crews patrol "the black" and do their best to put out all fires still burning. Root fires will burn underground, and if the weather gets hot enough, they can ignite again. The root can burn, and if the wind picks up, it can cause spot fires in areas that have not yet burned, or the fire can catch any unburned fuel on fire inside the black. Both scenarios can present problems when fighting the bigger fire. *Courtesy of the U.S. Forest Service.*

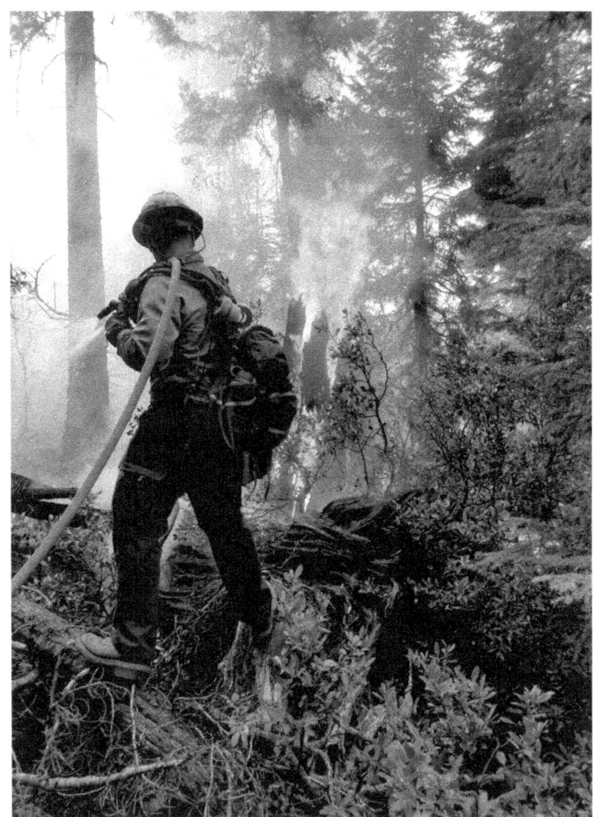

Right: A wildland firefighter using the hose on the Pelican Fire, 2017. *Courtesy of the Oregon Department of Forestry.*

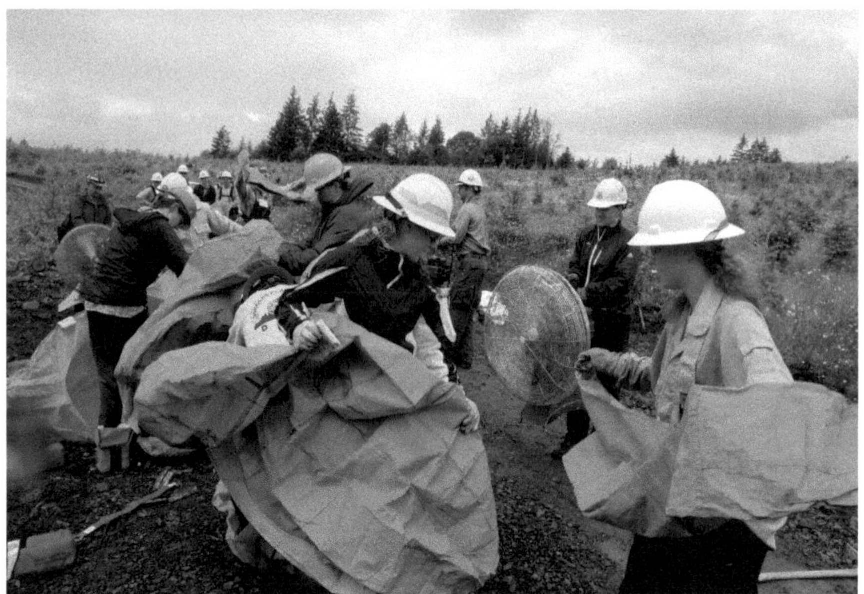

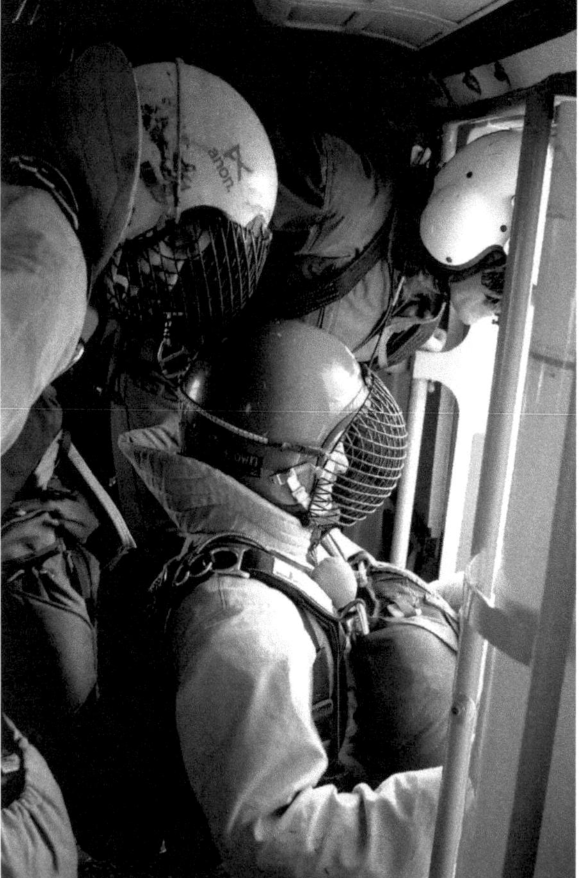

Above: The 2013 Mid-Willamette Valley Fire School students training with individual fire shelters. *Courtesy of Wikimedia Commons.*

Left: Smokejumper in the door on a training flight. *Courtesy of the U.S. Forest Service.*

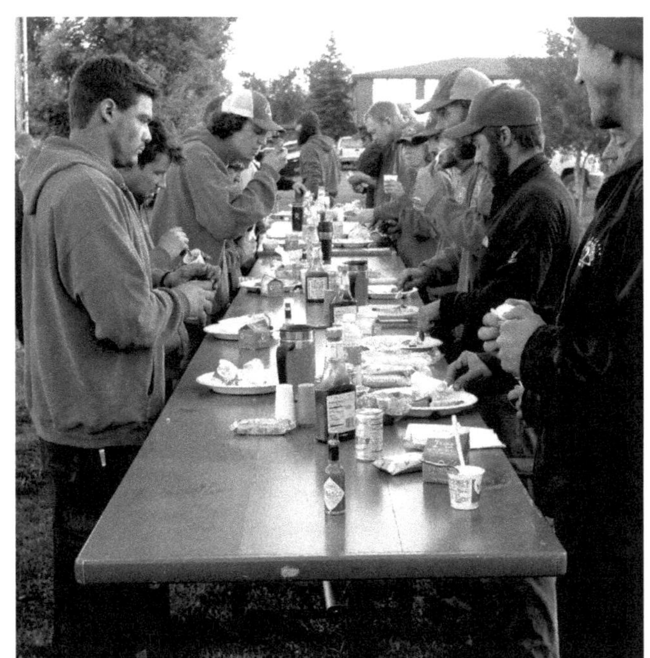

Wildland firefighters eating chow at the Bryant Fire, 2014. *Courtesy of the U.S. Forest Service.*

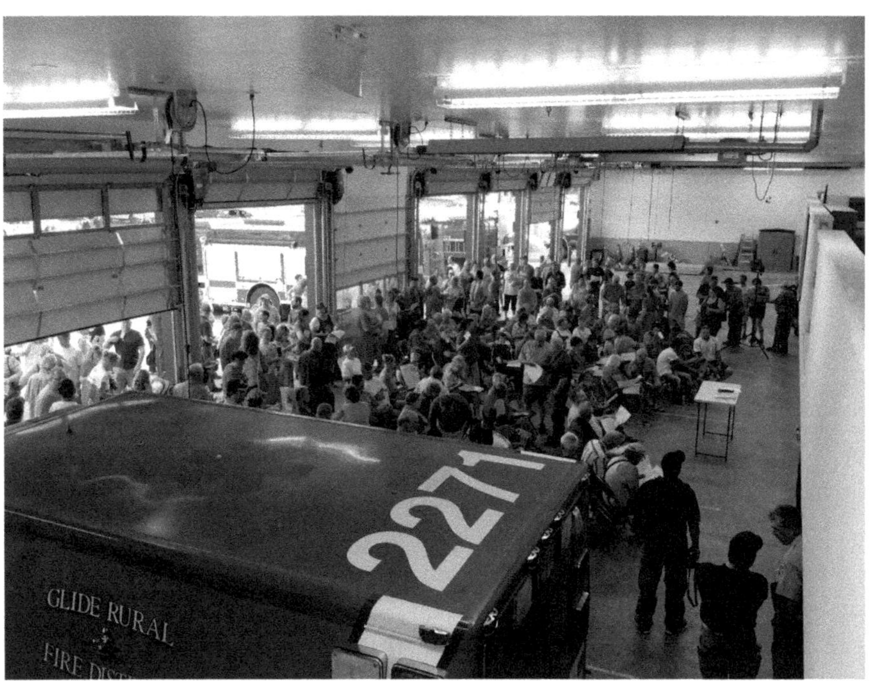

The leadership of the Cable Crossing Fire in 2015 held a community meeting at the Glide Fire Department to share information on the plan of attack with the public. *Courtesy of the U.S. Forest Service.*

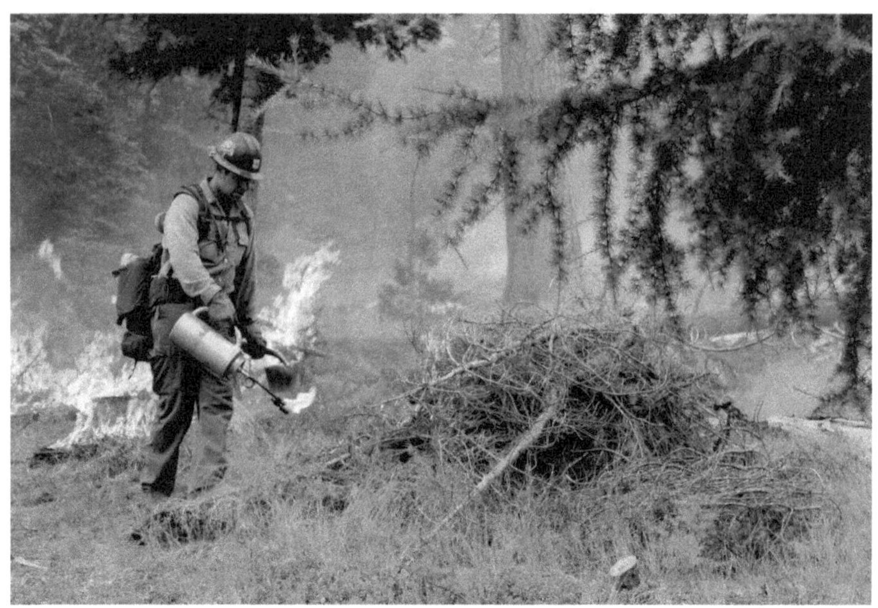

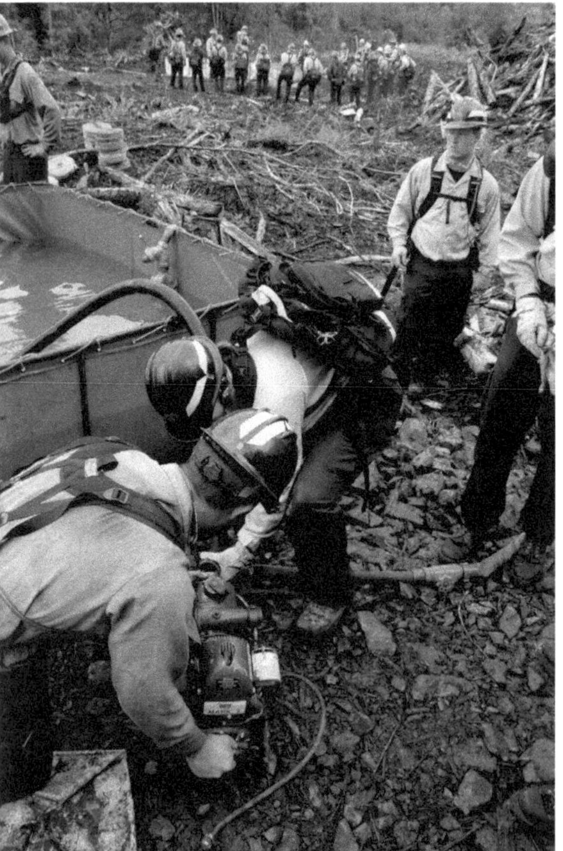

Above: A wildland firefighter using a drip torch to start a controlled burn at a location on the Canyon Creek Complex, 2015. *Courtesy of the U.S. Forest Service.*

Left: Pump use class at the Mid-Willamette Valley Fire School, 2013. *Courtesy of Wikimedia Commons.*

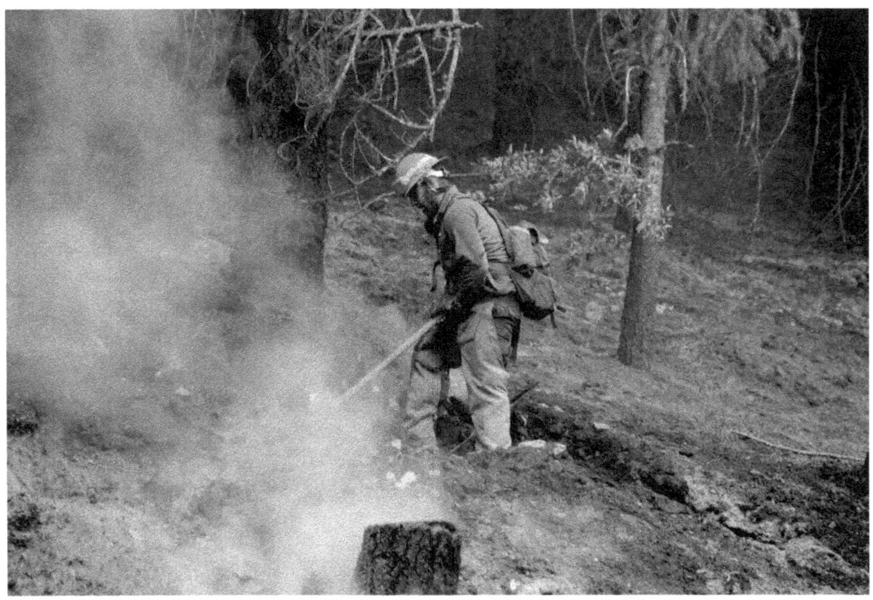

Wildland firefighter using a shovel to dry mop an area on the Horse Prairie Fire in 2017. *Courtesy of the U.S. Forest Service.*

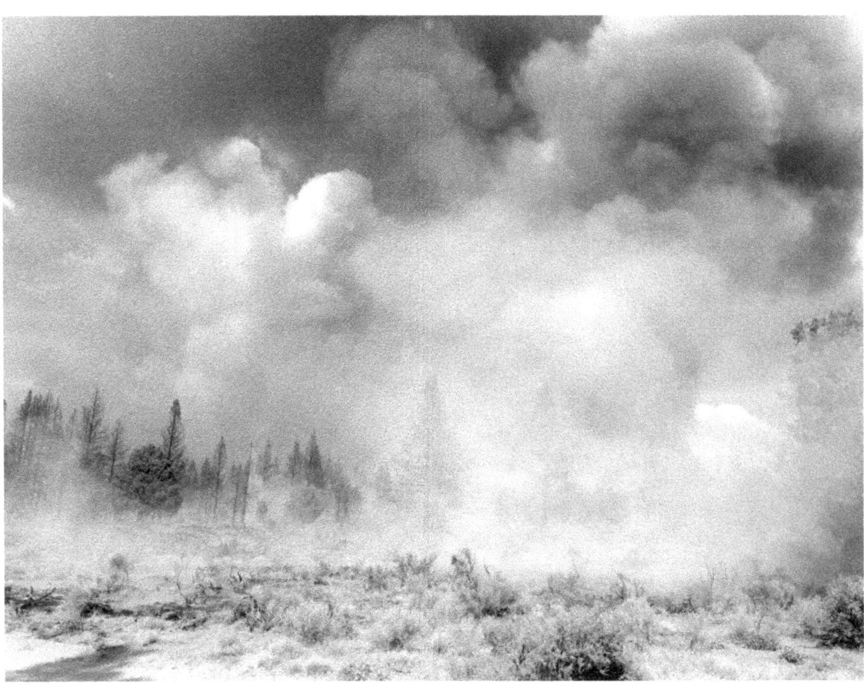

A wall of flame and smoke hundreds of feet high moving across the land as fast as the wind will carry it. This is the Moccasin Hill Fire in 2014. *Courtesy of the U.S. Forest Service.*

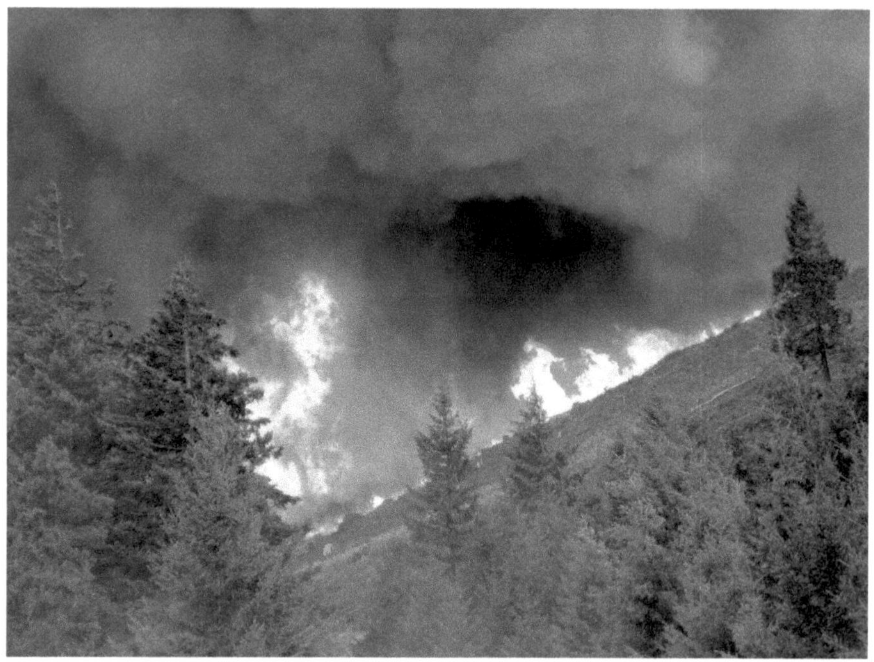

The Douglas Complex in 2013 was made up of two large fires that came together and ultimately burned 48,000 acres. The Douglas Complex ended up a Type I incident. *Courtesy of the U.S. Forest Service.*

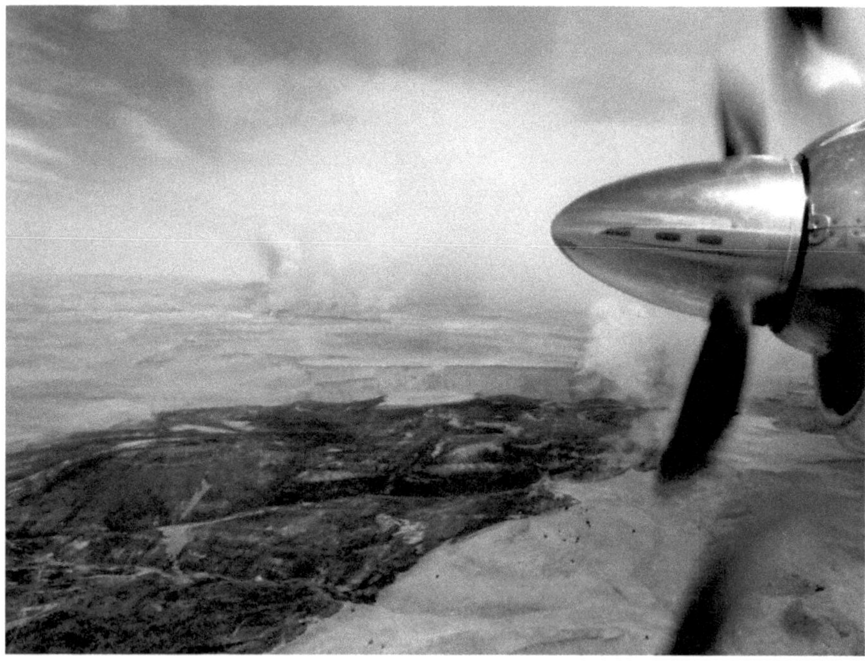

The leadership of the Buzzard Complex use a fixed-wing aircraft to scout the fire. The Buzzard Complex burnt over 100,000 acres and was about forty-five miles northeast of Burns in 2014. *Courtesy of the U.S. Forest Service.*

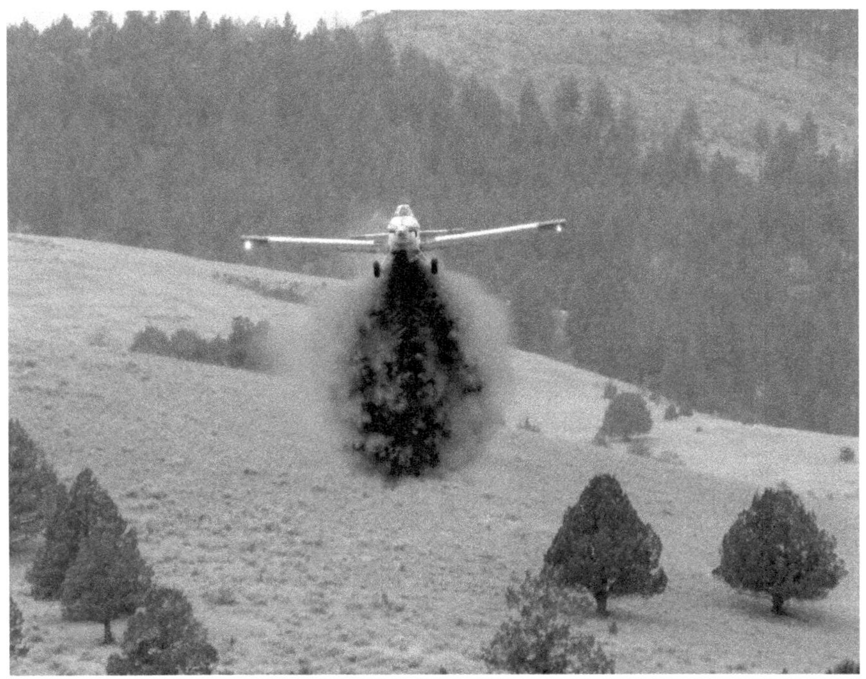

A fixed-wing plane drops retardant on the Government Flats Complex, 2013. *Courtesy of the U.S. Forest Service.*

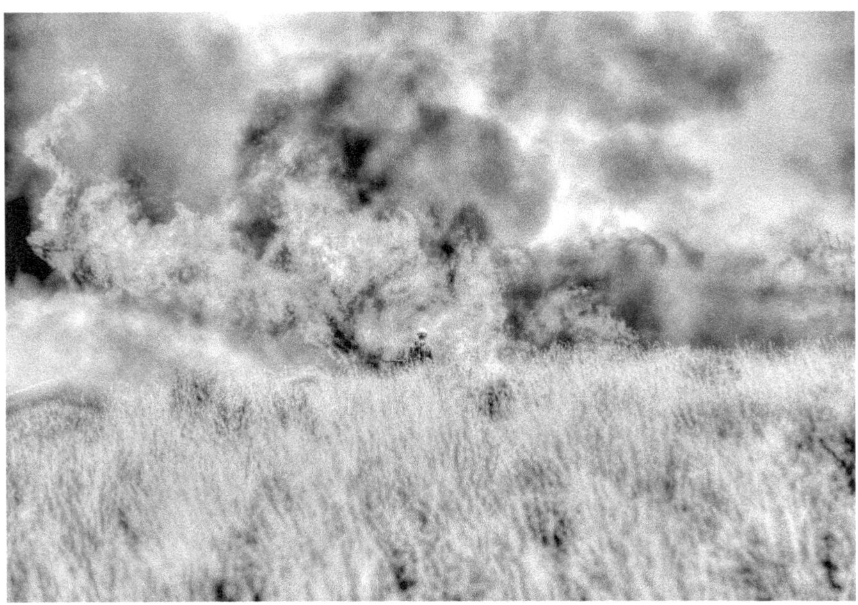

Just north of McDermitt, Oregon, a member of Fire Crew 7 keeps watch on the fire near the Blue Mountain Pass on the Long Draw Fire. This part of the fire was set by the crew in hopes of keeping the main part of the fire from interrupting the flow of traffic on Highway 95. *Photograph by Kevin Abel, Oregon BLM.*

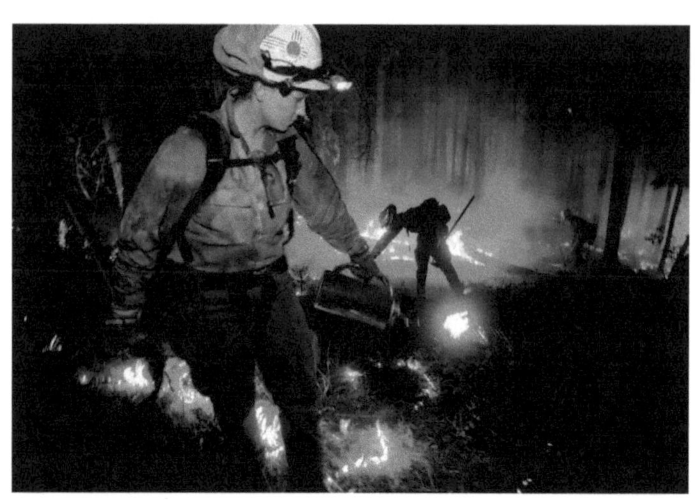

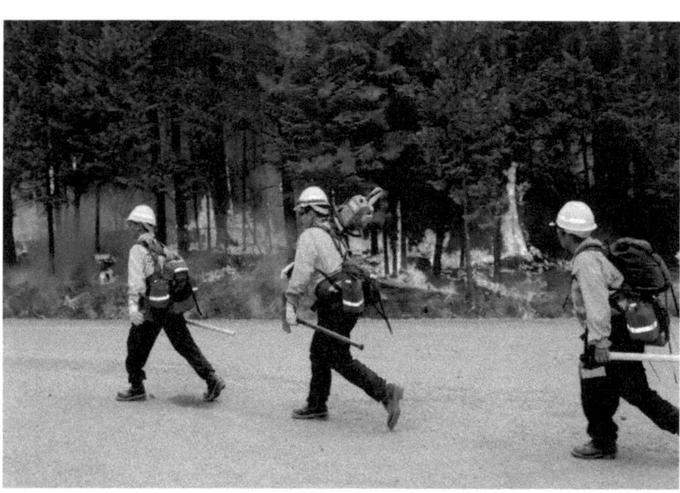

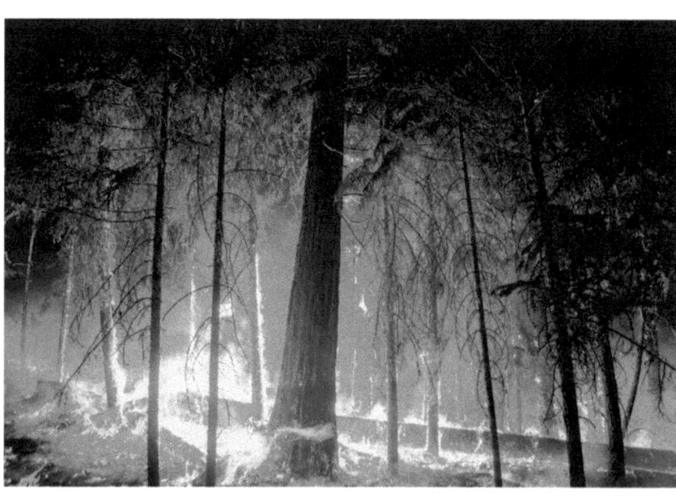

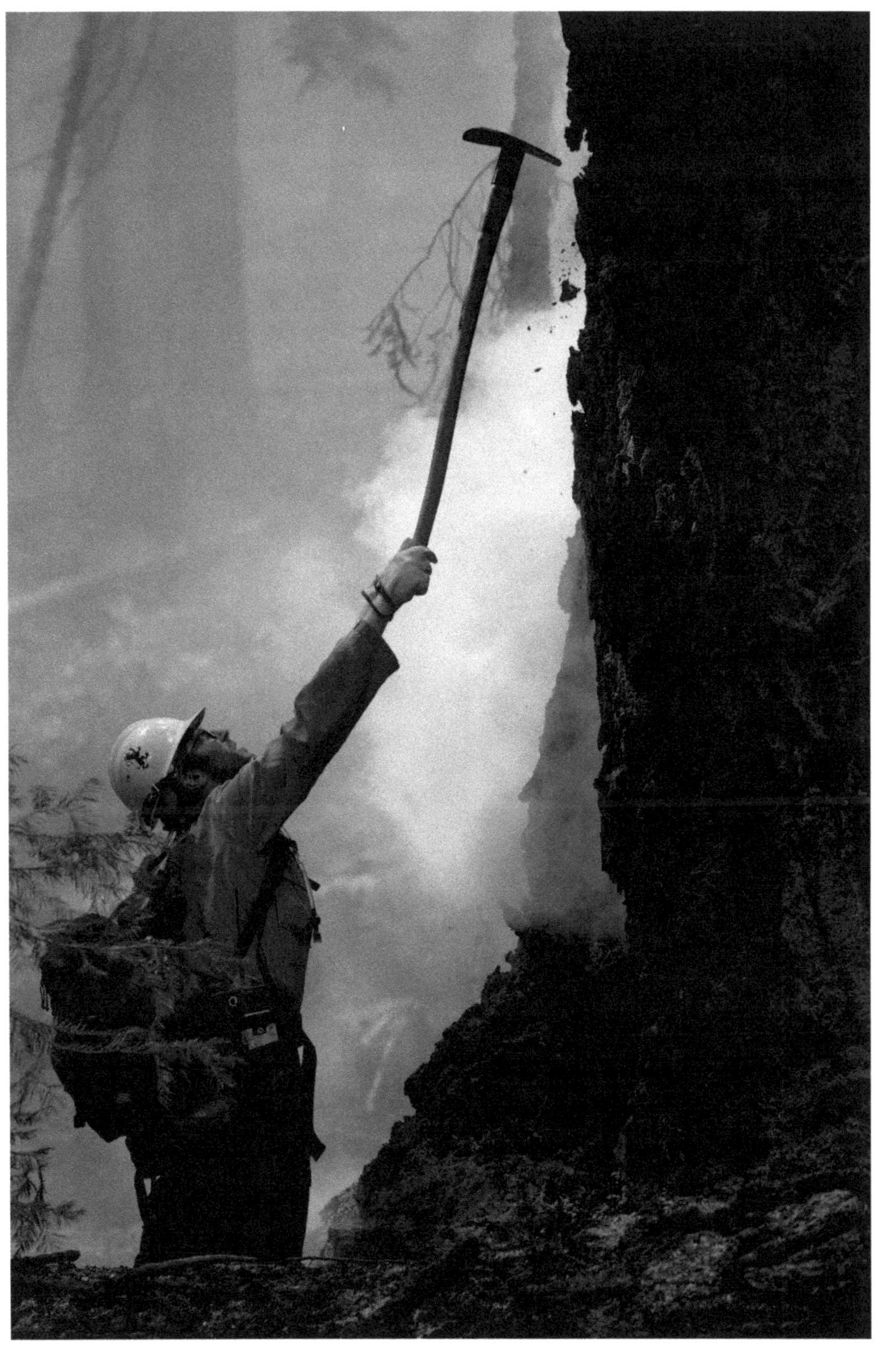

Above and opposite: Umpqua NF Fires, 2017, Oregon. *Courtesy of the U.S. Forest Service–Pacific Northwest Region.*

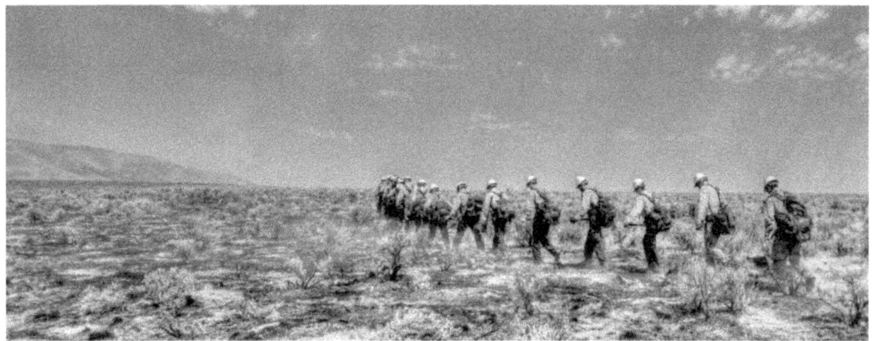

A hand crew marches to their area of operations. Men and women with tools in hand fought the largest fire Oregon has seen in over 150 years. The Long Draw Fire would go on to burn 558,198 acres of range land in Eastern Oregon. *Photograph by Kevin Abel, Oregon BLM.*

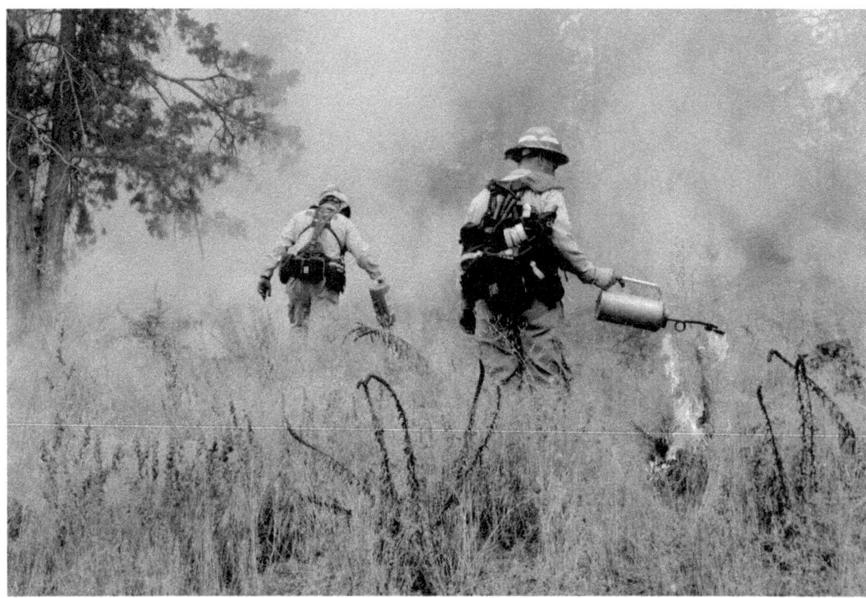

Two wildland firefighters use drip torches to intentionally burn an area to reduce the fuel on the Canyon Creek Fire, 2015. *Courtesy of the U.S. Forest Service.*

Chapter 6

FIGHTING FIRES IN OREGON, THEN AND NOW

The wildland firefighters haven't really changed over the years. Sure, the personal protective equipment has improved and firefighters have more tools today (such as air support and heavy equipment), and over the years, wildland fire agencies have learned many lessons to improve safety policies, but it still all boils down to men and women with tools in their hands scraping, cutting and spraying.

In the early 1930s, in order to battle the high unemployment rate and get people back to work after the Great Depression, the Emergency Conservation Work Act (ECW) and the Civilian Conservation Corps (CCC) were created by an executive order signed by Franklin D. Roosevelt as a part of his New Deal. The idea was to provide manual labor jobs while also conserving and developing rural lands owned by local, state and federal governments. Young, unmarried men could join, get trained and get paid. The Glacier CCC camp, the first one in Oregon, was established on Mount Hood, where Zigzag is today. Most of the participants were trained as wildland firefighters because that was the most pressing need.

Many of these men would go on to become leaders in the U.S. Forest Service (USFS) and, later, the Oregon Department of Forestry. A new wildland firefighter back then was taught what they are still taught today. Fire cannot burn without one of three things: heat, fuel or oxygen. This is called the fire triangle, and even still today, it is one of the first lessons an FFT II learns.

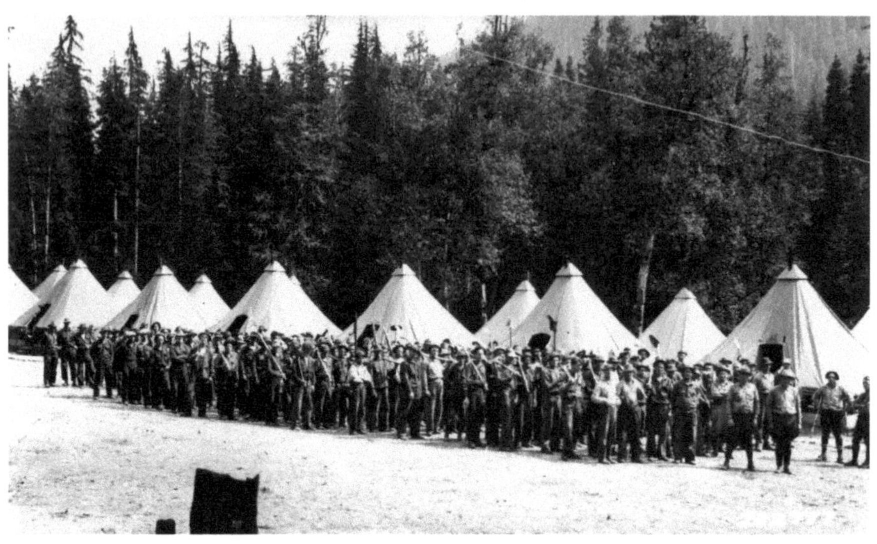

Forest firefighting training at Glacier CCC Camp, 1933. *Courtesy of the U.S. Forest Service.*

To fight fire, wildland firefighters do their best to remove one of the three elements in the fire triangle. Fuel is anything that burns in the forest, which is pretty much everything. Wildland firefighters work hard to prevent fires by removing underbrush and falling trees before fire season even begins. By doing this, the different agencies and people and companies that own forested land can try to prevent fires from growing should they start. Prevention can also range from thinning forests to putting out campfires and not smoking or lighting fireworks in the woods. The USFS has two mascots that do their best to prevent wildland fires: Smokey Bear and Woodsy Owl.

If prevention fails and wildland firefighters are on a fire, the next thing they will do is remove the fuel in other ways. Wildland firefighters in Oregon are taught to be aggressive when preventing the spread of fire, and the best way to do this is to dig a fire line all the way around a small fire or anchor a fire line to a something on the perimeter that will not burn. This can be a road, a wide creek, a river, a dozer line or something similar. Digging a fire line means taking an axe, a pulaski, a shovel or any other tool that can remove vegetation and scrape a wide line all the way down to mineral soil. The width of the line depends on the weather, terrain and other factors that a crew boss or strike team leader will evaluate.

Another way to take the fuel out of the equation is a controlled burn. Wildland firefighters use this technique on just about every large fire. To

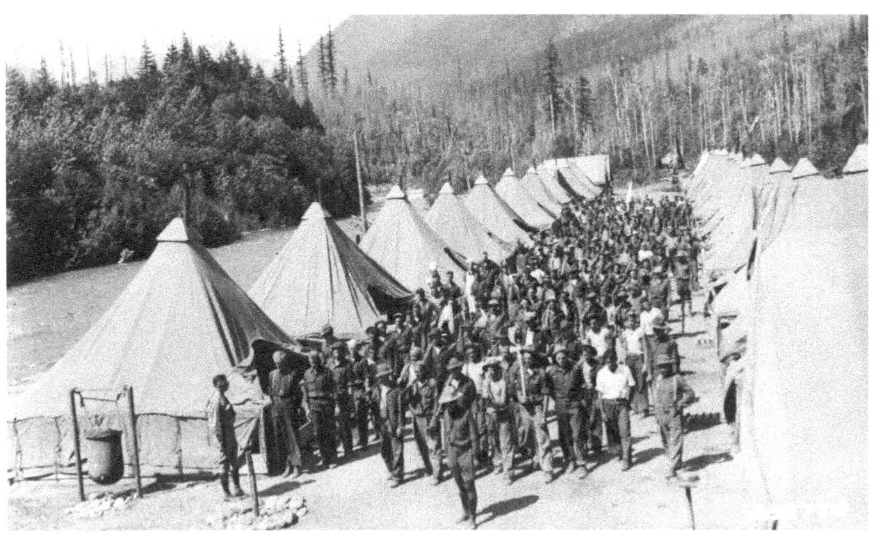

Forest firefighting training at Skagit CCC Camp in Mount Baker National Forest, 1933. *Photograph by R.L. Fromme, courtesy of the U.S. Forest Service.*

the layman, it may not make a whole lot of sense to fight a fire by setting a fire. If looked at objectively, a person wouldn't be blamed if they asked the question, "Doesn't that just make a larger fire?" In a way, it does, but the intent is to burn away the fuel under conditions that allow a crew to only burn what they want to burn. So, if an incident commander sees a weeklong weather forecast, the surrounding terrain, the predicted relative humidity and other factors, he or she may determine in which direction the fire will burn. If he or she has the division burn between the larger oncoming fire and a road or dozer line, by the time the fire reaches that area, all the fuel will be gone. In this way, the incident commander (whether it be for a Type 1 giant fire or a Type 5 small fire) can control the direction and intensity of the fire.

There are other reasons for controlled burns. They improve the forests' overall health and help to reduce the chances of a larger fire. The USFS describes controlled burns as "nature's way of recycling."[16] The forest floor is littered with branches and snags left to rot. Again, according to the USFS, large logs can take one hundred years to decompose. Pine needles and cones decompose slowly, too. "It takes more than a year for pine needles to decay. As a result, year after year, pine needles continue to build up until they are eliminated by fire." A controlled burn helps with this process, but there are many places in Oregon where controlled burning will not work.

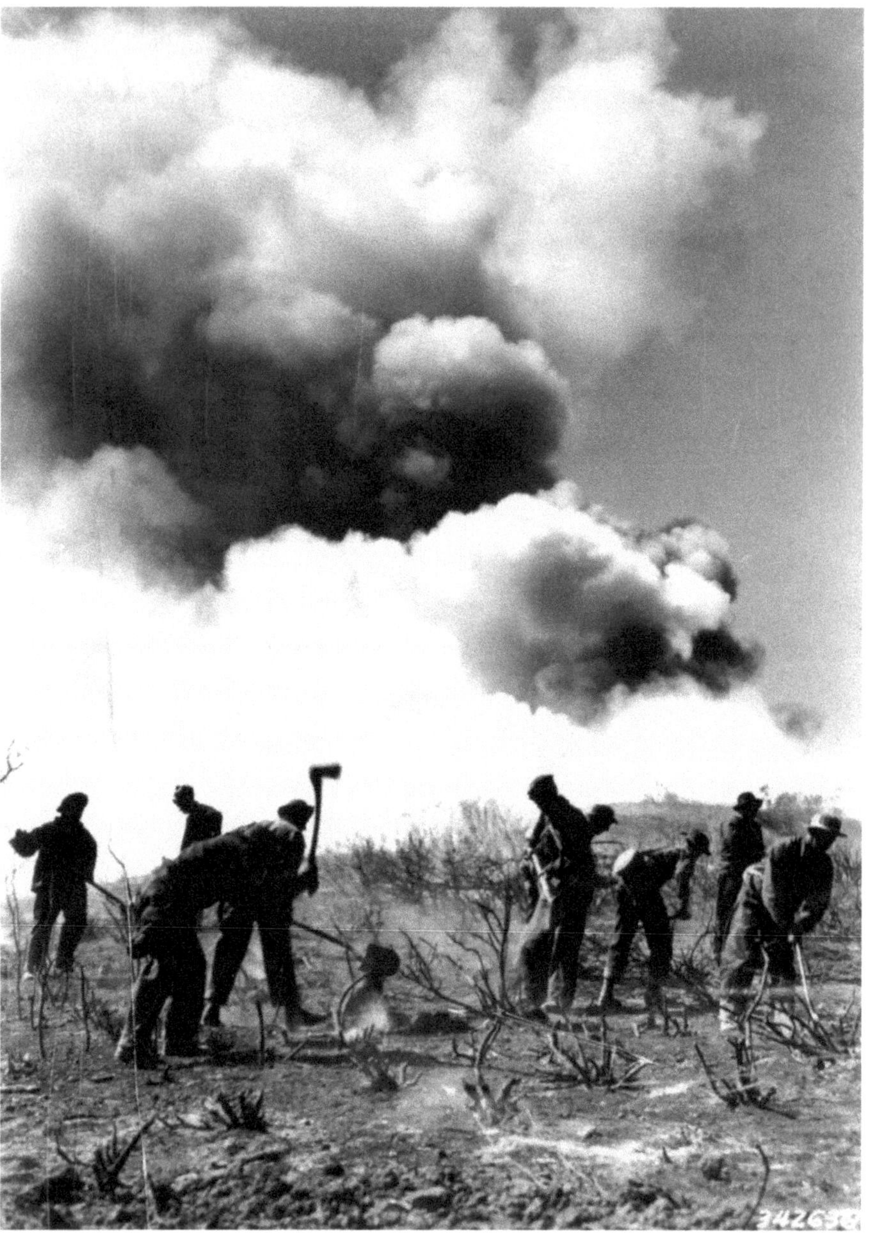

A crew from the CCC cuts a fire line in front of a large smoke plume in Eastern Oregon, 1935. *Photograph by F.E. Dunham, courtesy of the Gerald W. Williams collection at Oregon State University.*

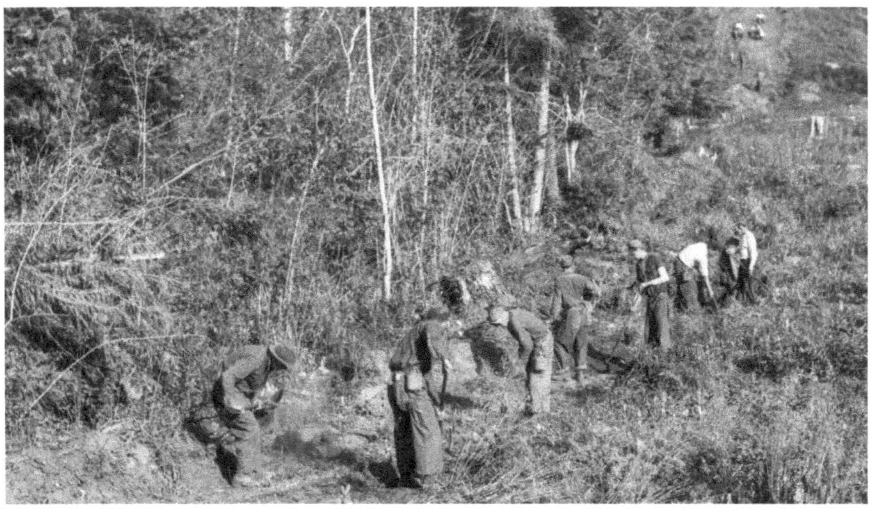

North Bend CCC camp crew digging a fire line in the Siskiyou National Forest, October 19, 1936. *Photograph by Ernest Lindsay, courtesy of the Gerald W. Williams Collection at Oregon State University.*

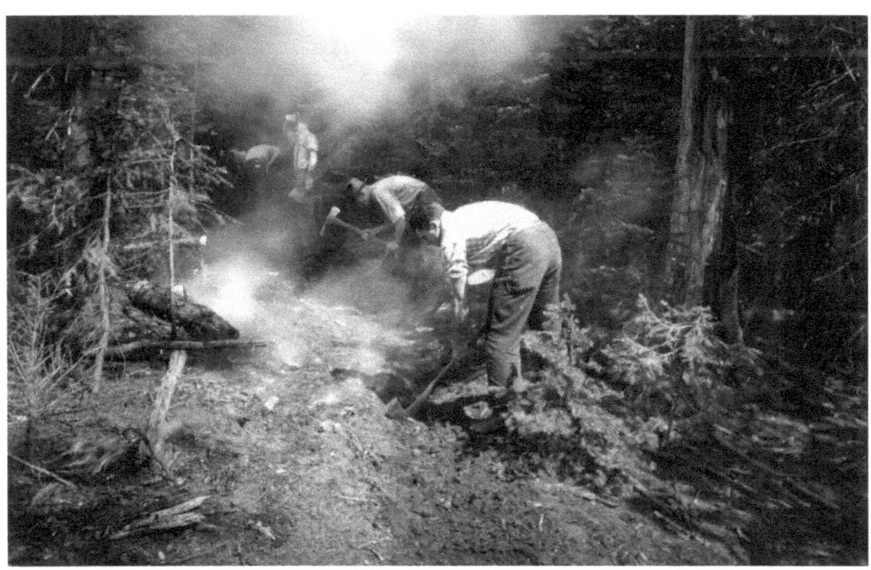

Wildland firefighting instruction at Guard Training School in Tollgate, Oregon, in Umatilla National Forest in June 1939. *Photograph by Ray Filloon, courtesy of the U.S. Forest Service.*

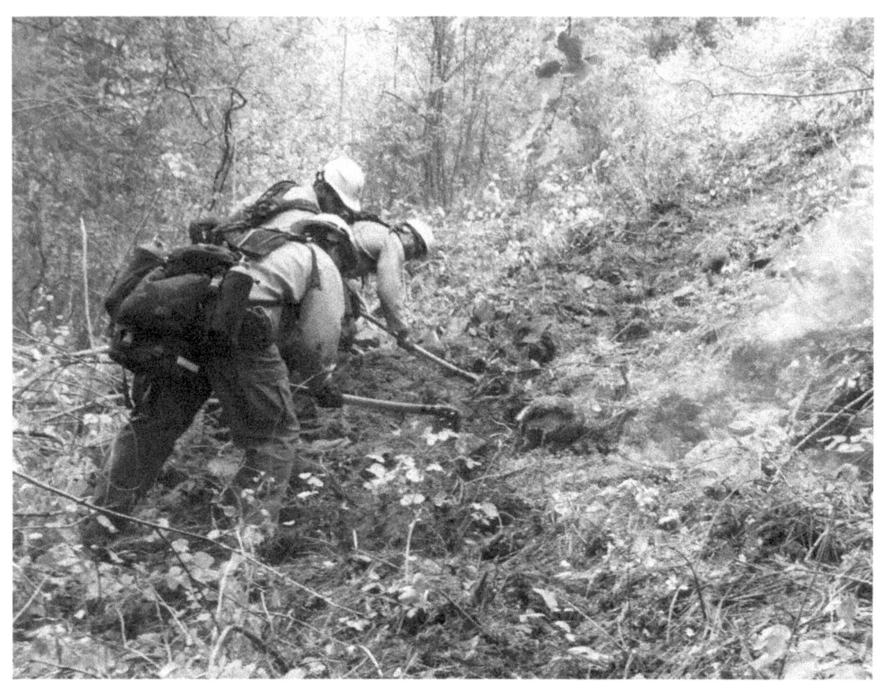

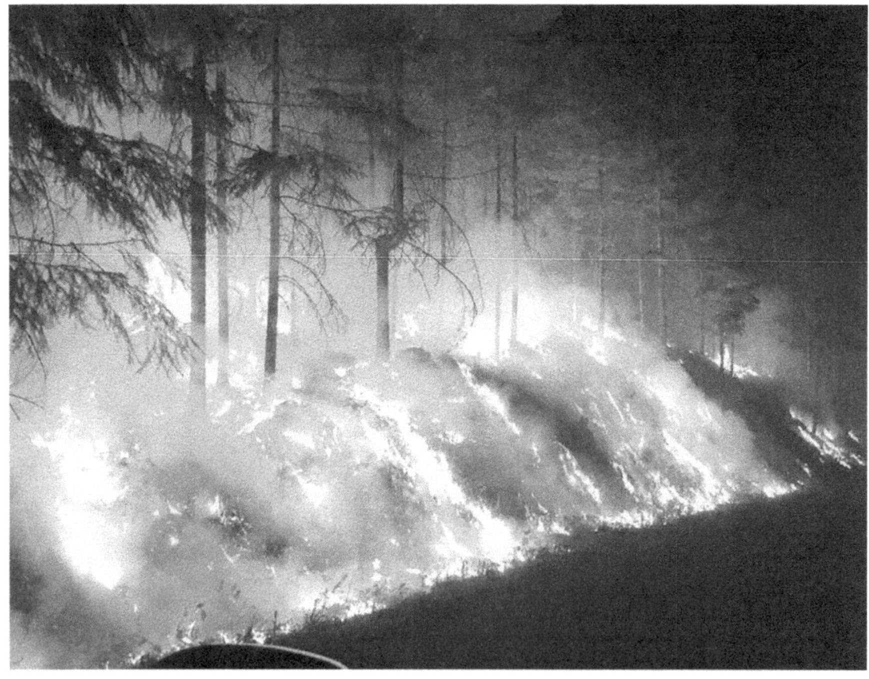

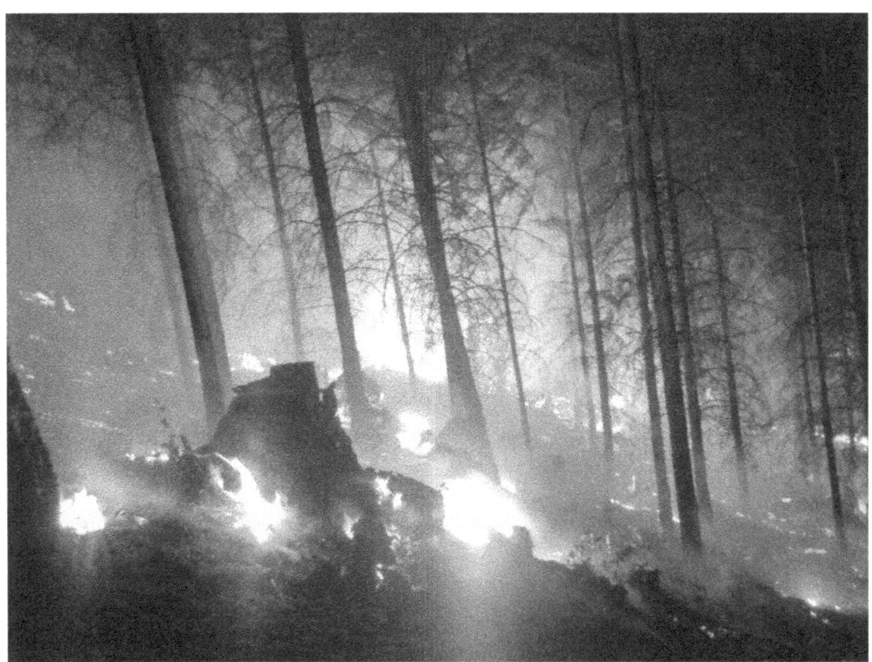

Above: Notice the low intensity of this burn. This means it was done under the correct conditions. When the fire is "skunking around" or "crawling," the chances of it jumping a road, dozer line or fire line are low. This was on the Stouts Fire in 2015. *Courtesy of the U.S. Forest Service.*

Right: Tillamook Fire blows up, August 25, 1933. The smoke column rose eight miles high and was visible throughout western Oregon and Washington. *Courtesy of* The Oregon Encyclopedia.

Opposite, top: Hand crew digging a fire line in 2017. *Courtesy of the U.S. Forest Service.*

Opposite, bottom: A controlled burn on the Happy Dog Fire outside of Glide, 2017. *Photograph by the author.*

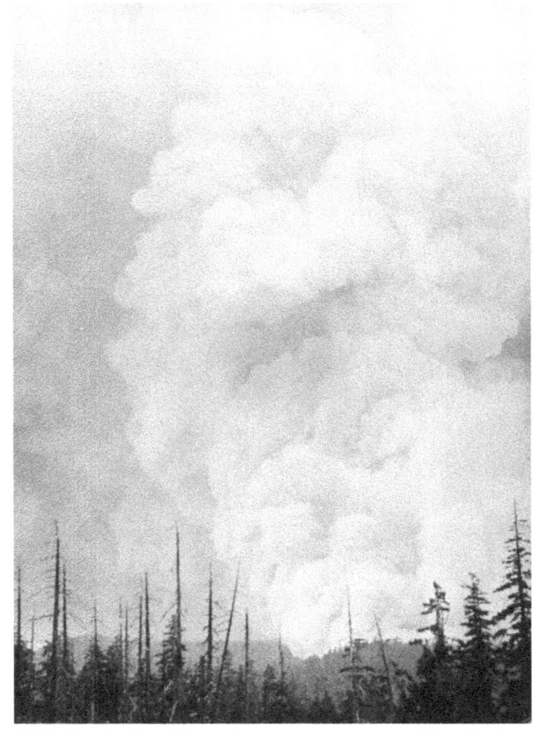

Oregon is beautiful and full of many different types of regions and climates. Each one presents unique problems when fighting wildland fire. For example, the Oregon coast is populated with old-growth cedars, sitka spruce, Douglas firs and western hemlock with a dense understory of rhododendron, salmonberry, black twinberry and wax myrtle.[17] Add the rocky terrain and coastal winds, and it's no wonder some of the state's largest fires occur in the Tillamook State Forest. The eastern part of the state has its own hazards. Strong winds and the 2.2 million acres of juniper forest in Eastern Oregon present a whole different set of problems. In the chapters to come, this book will chronicle some of the larger fires in the state, but for now, let us look at how our wildland firefighters fight these fires.

Fire prevention is very important, but no matter how much we try to prevent wildland fires, they will happen in Oregon. Lightning strikes, sparks from train tracks, heavy machinery working in the woods—there are so many different ways a fire can start. Once they do, wildland firefighters are called to attend to them. Other than removing fuel, what can a wildland firefighter do? They can reduce the heat by spraying the flames with water.

We covered the different sizes of engines earlier, but what you need to remember about water is the more gallons you have, the more you sacrifice mobility or access to the fire. Wildland firefighters have to be very creative and stingy with their water. Any wildland firefighter can carry a five-gallon bladder bag on his or her back. A Type 6 engine is small and can usually maneuver on the logging roads and even dozer lines, but a Type 4 engine may not be able to handle these same obstacles.

The best way to get water to a fire in many cases is via a portable tank, a pump and a hose lay. A hose lay is when a crew deploys hundreds or maybe even thousands of feet of hose. The largest hose would be the trunk line, and every two hundred feet, a "gated wye" and a reducer are used to connect a one-hundred-foot-long smaller hose. The trunk line for wildland firefighters is the largest hose in their inventory, which is an inch and a half in diameter. If you look at a structure fire crew that works for the county or city, they can hook up a two-and-a-half-inch hose, and with maximum pressure, that hose can pump eight hundred gallons per minute. Type 5 and Type 6 engines would be dry in sixty seconds, and the fire would continue raging. When putting down a hose lay, hundreds of feet of hose are carefully placed around a fire with a "gated wye" every two hundred feet. This gated wye splits the trunk line in two. One side continues the trunk line, and the other is fit with a reducer to attach a hose with a smaller diameter (one inch). The end of the smaller hose has a forester nozzle for direct spray and mist.

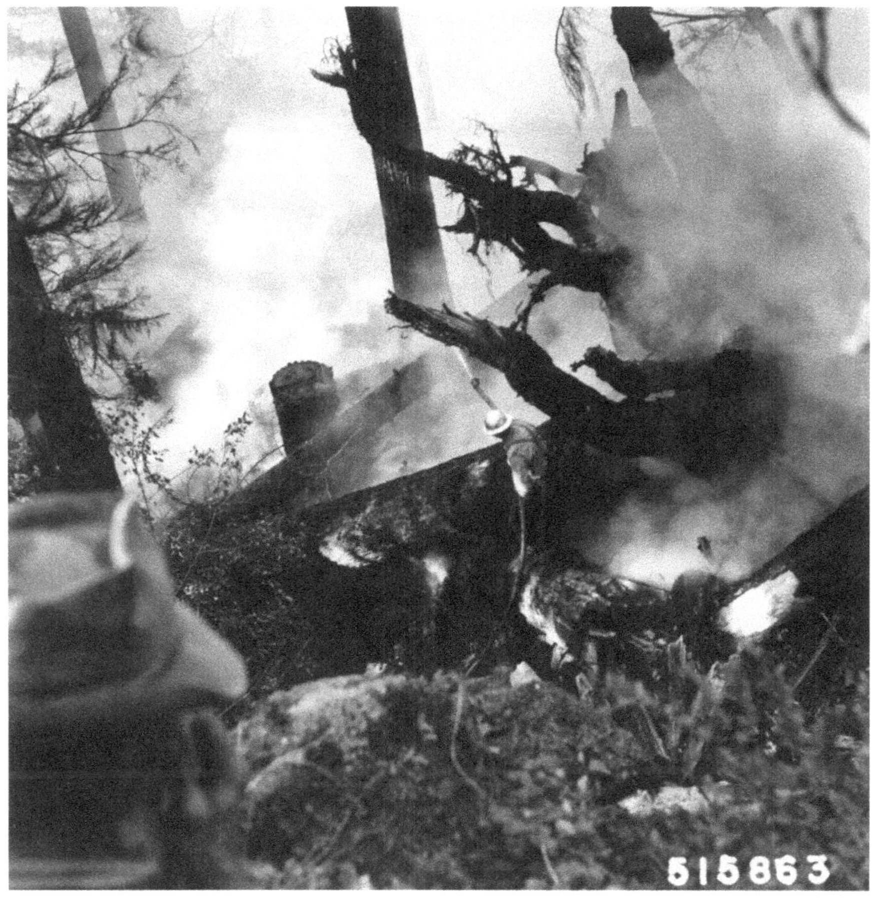

A logger hastily recruited to help combat an escaped slash fire burning green Douglas fir timber. He braved the heat and smoke to get water on the fire in September 1966. *Photograph by Samuel T. Frear, courtesy of the U.S. Forest Service.*

Any of these smaller hoses may be needed at any time. This means that the hose lay needs to have water available. Every wildland firefighter must conserve water for the greater good.

Another reason wildland fire hoses are smaller in diameter than structural fire hoses is because wildland fire engines don't have room for larger and more powerful pumps, but again, more powerful pumps mean more gallons per minute. Many Type 6 engines only have three hundred gallons of water and maybe some fire retardant foam. When putting out a hose lay, a wildland firefighter must think about the amount of water in the tank, the size of the pump pushing the water, the diameter of the hose, how many feet of hose

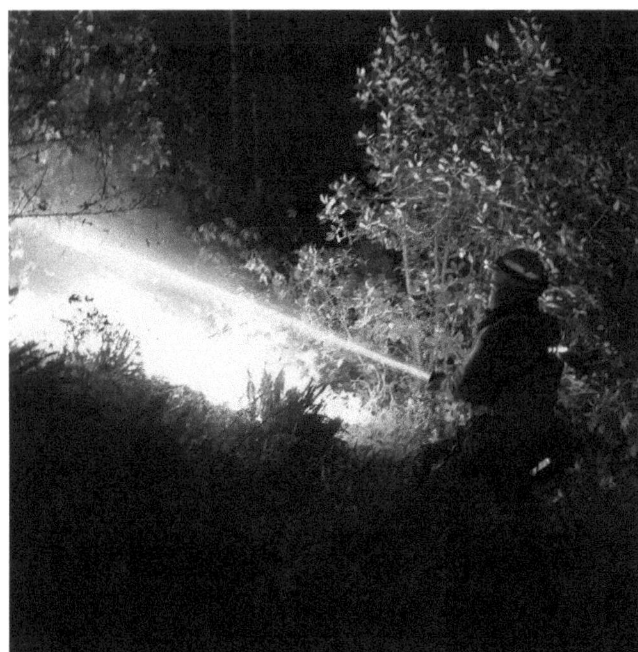

Wildland firefighter Julio Molina "putting the wet stuff on the hot stuff" on the North Umpqua Complex, fifty miles east of Roseburg, in 2017. *Courtesy of Dacoda Davis.*

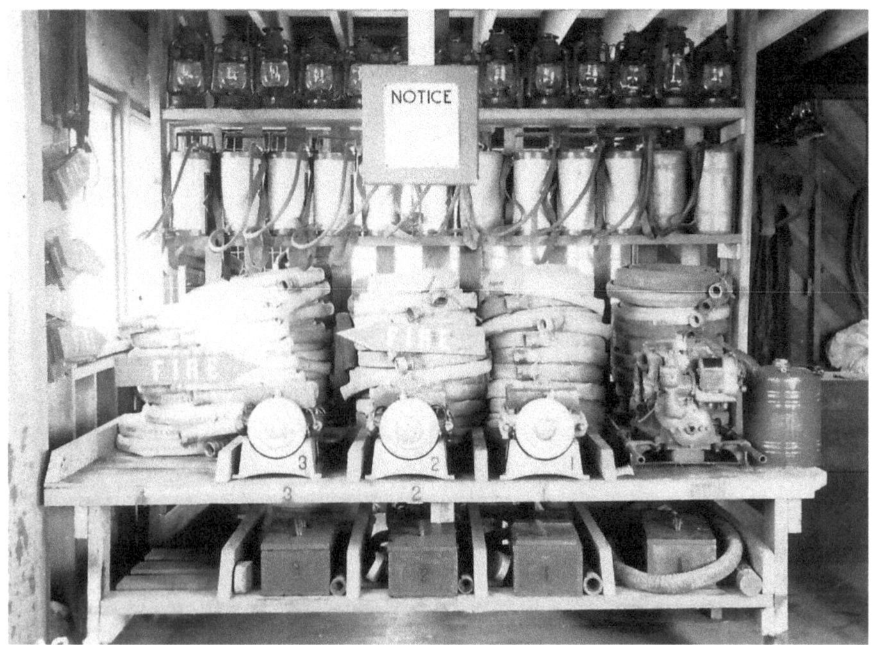

Fire equipment ready to go in a fire cache building—hoses, hand pumps and kerosene lamps, 1937. *Courtesy of Forest History Center/Oregon Department of Forestry.*

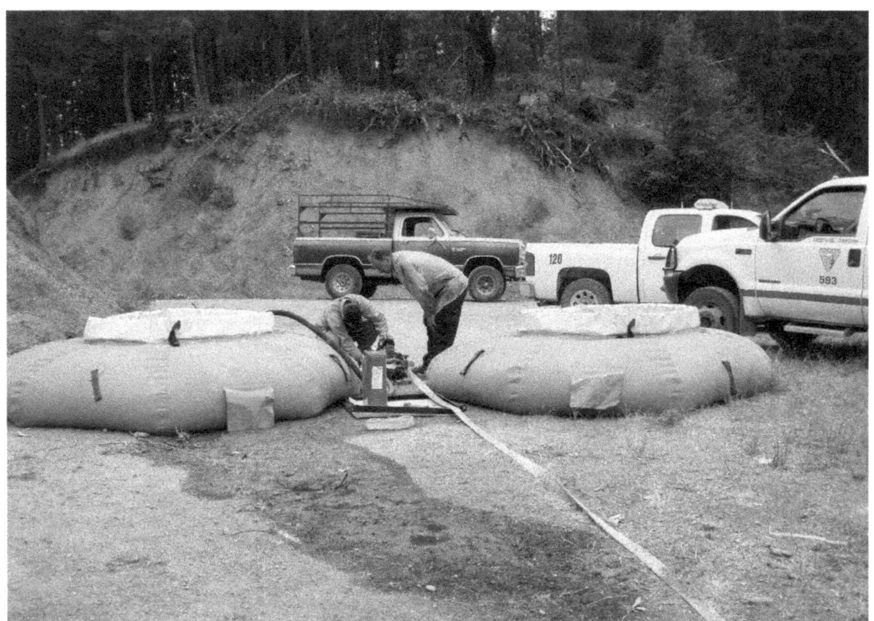

Wildland firefighters filling up pumpkins on the Brimstone Fire, 2013. This was a fire headed up by the Oregon Department of Forestry. The fire was estimated at 2,300 acres and manned by approximately seven hundred personnel. *Courtesy of the Oregon Department of Forestry.*

there are and the terrain. Even the pump on a Type 4 engine will have a very difficult time pushing water a couple hundred feet up a steep hill. The bottom line is that whether a wildland firefighter is using the trunk line from a Type 4 engine or a five-gallon bladder bag, he or she must be creative, conservative and aware of how much they're using for any fire.

Oregon has many rivers, creeks, lakes and ponds from which a wildland engine can draw water, but sometimes that is not enough. Luckily, wildland firefighting leadership teams can deploy self-supporting open-top water tanks, also called "pumpkins." These water tanks are able to be folded up and come in a variety of sizes, the largest being fifteen thousand gallons.

The last part of the fire triangle is oxygen. How do you take oxygen from a fire? Water is a luxury most the time for hand crews and hotshots. Most crews have backpacks with five-gallon bladder bags and a pump hose. Five gallons is hardly enough water to do much, but a good wildland firefighter is creative and resourceful. Five-gallon water backpacks have been part of the wildland firefighting tools for a long time.

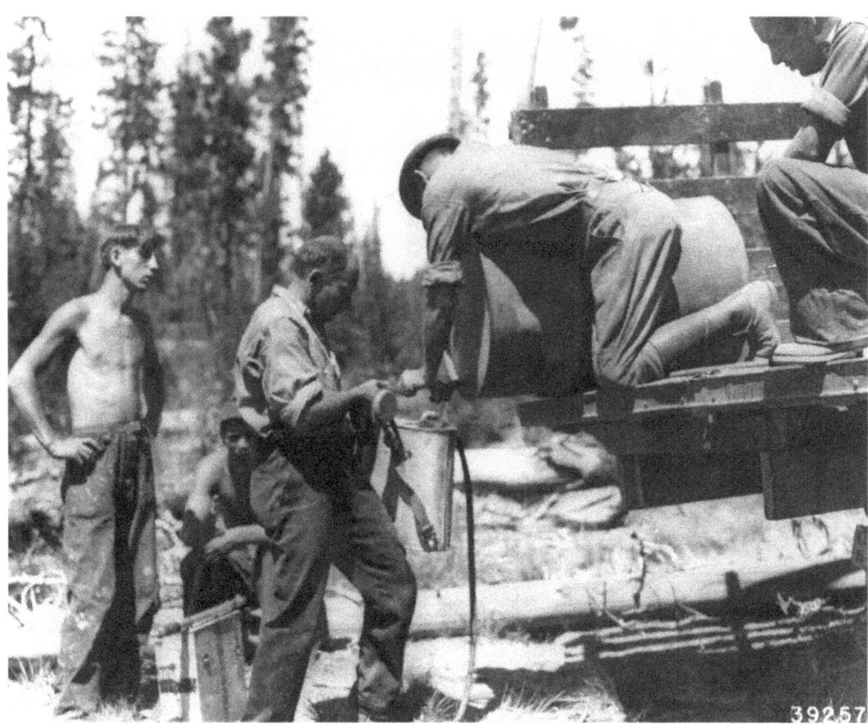

Above: Wildland firefighters filling backpack sprayers at the Mokst Butte Fire in the Deschutes National Forest, 1936. *Photograph by Ernest Lindsay, courtesy of the Gerald W. Williams Collection at Oregon State University.*

Right: A shirtless wildland firefighter wets down a log on the Mokst Butte Fire, 1936. *Courtesy of the Gerald W. Williams Collection at Oregon State University.*

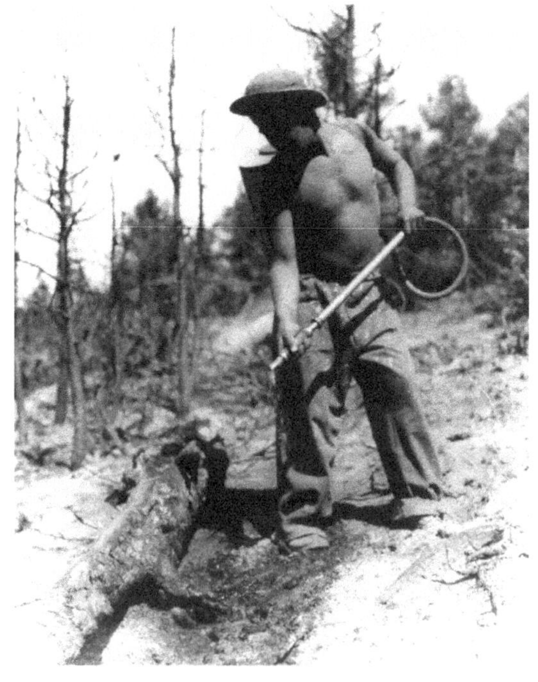

Throughout history, wildland firefighters have come up with many different ways to fight fire. It was common practice decades ago to fight fires by smothering them with a wet burlap bag. Burlap bags are not required today, but some engines still carry them. A more common practice is using tools to break apart burning logs and stumps and then trying to smother the burning wood with dirt. This is a commonly known as dry mopping.

It seems pretty easy to understand that with a tool in hand, a wildland firefighter can smother a fire or scrape a line to keep the fire from spreading. A wildland firefighter on the ground will fight fire with tools or with water, and put that way, it seems to make perfect sense. It's simple enough that it seems like anyone can do it, but let's look at some of these fires.

It takes a brave soul to get out there and face that danger every day. Not only do wildland firefighters face that danger, they often work until they are exhausted. The bigger the fire, the more the danger and the longer the shifts. Imagine getting up at the break of dawn, getting your assignment and heading out to fight fire with nothing but a tool or maybe a hose with

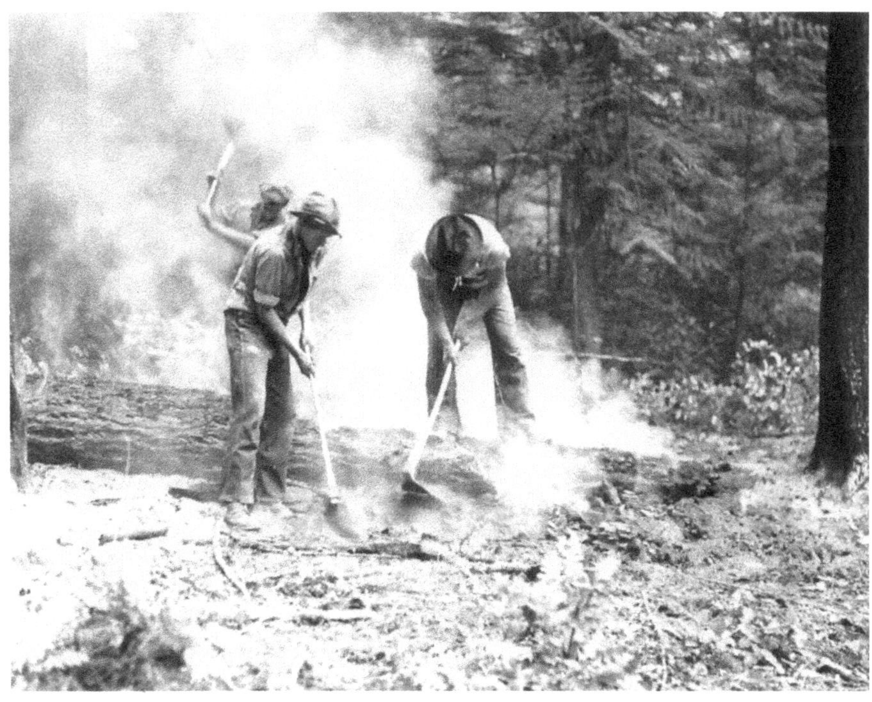

A group of wildland firefighters cool down a log, 1937. *Courtesy of the U.S. Forest Service Historical Archives.*

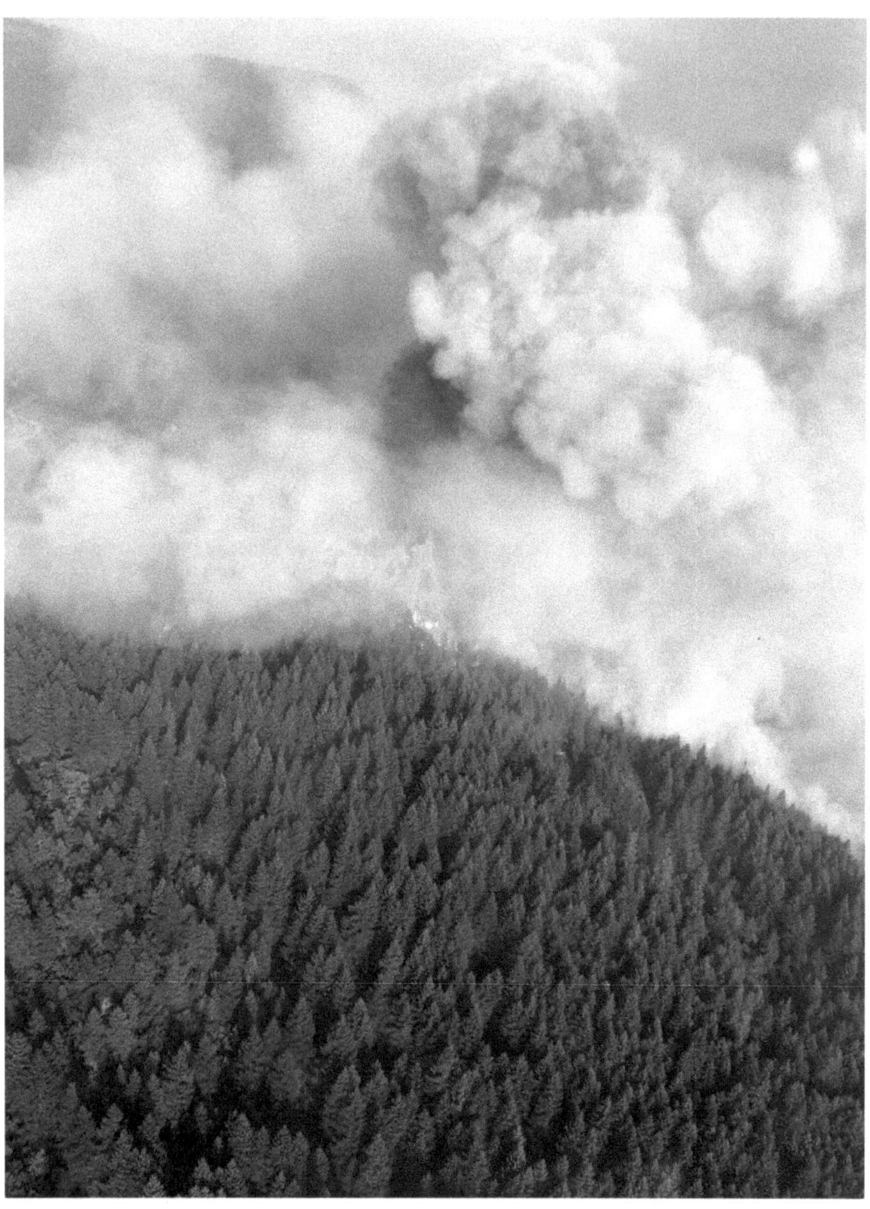

The Horse Prairie Fire was a Type 3 incident that burned 16,400 acres near Canyonville in Douglas County, 2017. *Courtesy of the U.S. Forest Service.*

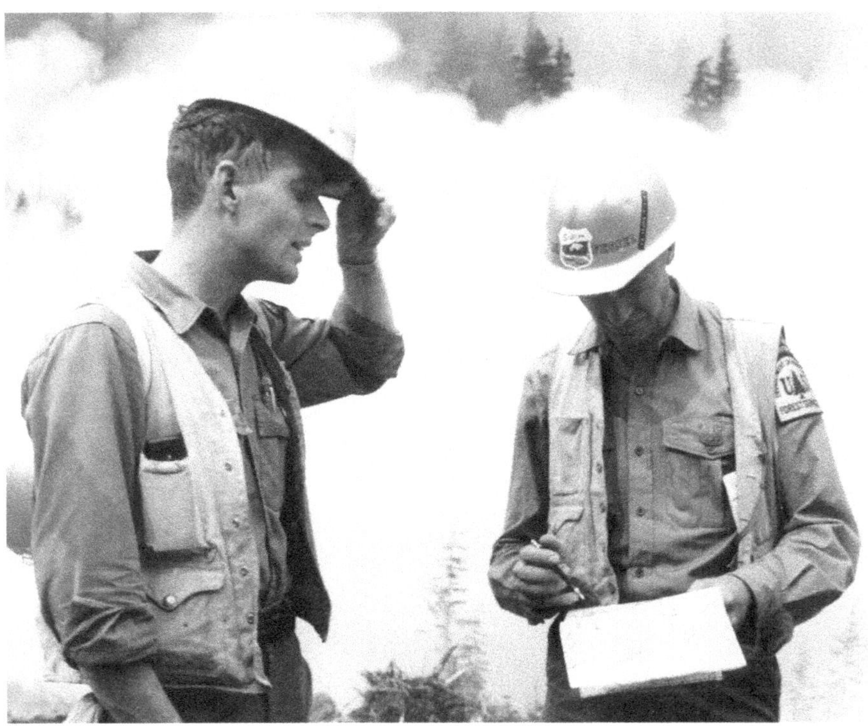

Two Siuslaw National Forest "overhead" discuss tactics at the Buck Mountain Fire in the Detroit Ranger District in Lane County, 1967. *Photograph by Sam Frear, courtesy of the Gerald W. Williams Collection at Oregon State University.*

a limited amount of water (if you're lucky). Do that for sixteen hours, with very limited breaks, and then come back, eat, maybe take a shower and get a few hours of sleep before doing it all over again. Do this for two weeks straight, take forty-eight hours off, and do it again. This is what wildland firefighters do all season. The leadership, or overhead, spent most of their careers on the line fighting fires. Many of them try to get out to see the fire if they can. They may not do the physical labor as much as they did before, but they do have more responsibility. Many of them get less sleep than the men and women on the line because they have to plan, go to meetings with other leadership and the public and ensure their division, strike team, area and so forth have the resources they need to put out the fire.

There are other jobs in wildland firefighting besides fighting fires. There are fire lookouts, slash and burn crews and even rehabilitation teams that continue working long after the fire is over. State and federal agencies

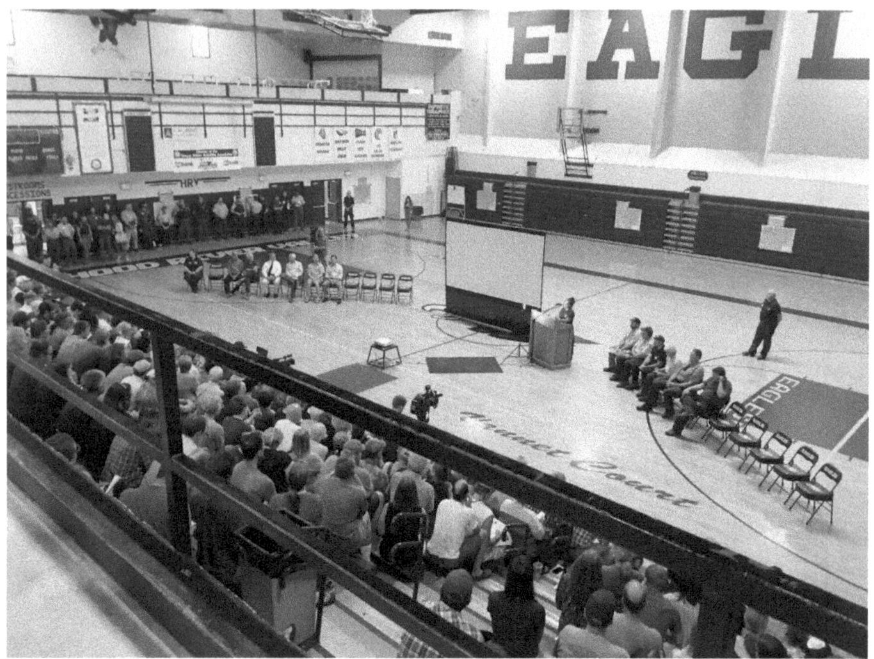

The leadership, or "overhead," having a public meeting to give the public news of how the fire will spread and approximately how long it will take to get to 100 percent under control on the Eagle Creek Fire in the Columbia Gorge, 2017. *Courtesy of the U.S. Forest Service.*

push these types of jobs to the winter for a couple of reasons. The first is because in the winter, a controlled burn has a very low chance of growing out of the designated area. Another reason is to give the permanent or year-round employees work outside of preparing for the next season. Some of the year-round employees join a county or city's fuel mitigation crew—a team of wildland firefighters whose daily job during the off-season is "fuels reduction" work. This means thinning vegetation and piling what they cut, maybe covering it with a tarp and burning the piles when weather conditions lower the chances of an intense fire. Fuel mitigation crews are also known as slash and burn crews. A fuel mitigation crew may prepare locations that have a high probability of burning in the coming season.

Seasonal wildland firefighters go back to their full-time jobs or try to find something to do in the off-season. There are some wildland-related jobs they can do. From the end of winter through early summer, a few may find jobs as fire lookouts. There are not as many as there used to be, but it's good work if you can get it.

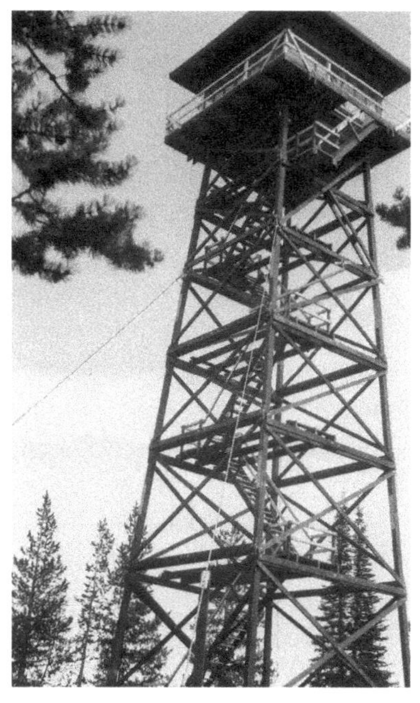

Desolation Butte Lookout Station, Umatilla National Forest, 1963. *Courtesy of the U.S. Forest Service.*

Fires have been reported by people living in fire lookouts for over one hundred years. The lookouts are high places with great views where one or two people would live for months. Their only job is to watch for smoke in the forests and call it in if they see any. Fire lookouts were built all over Oregon, but many have been closed down because of safety concerns and other reasons. A lot of the lookout stations were built decades ago, and it's just not cost-effective to rebuild them and add today's safety features, so they are taken down. Also, cameras can do the job more efficiently and cheaply than paying someone to live in remote areas. Today, only a fraction of lookouts are manned and operable, but dozens of old lookouts are available to rent for recreational use through the USFS.[18]

Another way to spot fires in the forests and wilderness areas is to see it from the air. Wildland firefighting organizations will send out an aerial patrol when they believe there is a high chance of a fire starting, like after a thunderstorm, but they are not the only pilots who report fires. There has been an increase in commercial flights coming to Portland and smaller airports in the state, and if any of them see a fire or a smoke column, they will report it. Commercial pilots will notify the Oregon Department of Forestry or the USFS depending on where the fire is located.

Of course, aircraft are also an offensive tool when fighting fires. Helicopters and planes are used for dropping water and retardant on fires. Most of the rotary and fixed-wing aircraft used for water and retardant drops in Oregon are hired via contractors. Helicopters, or rotary-wing aircraft, can drop water and spray retardant and can be used to blacken an area in a controlled burn in order to reduce the fuel so a fire can't spread. Of course, helicopters can move more slowly than planes and hover. This makes them a valued resource on a fire. A wildland firefighter on the ground can call for water near their position, and a pilot may be able to see enough—depending on the smoke—to make the drop as accurate as possible, but helicopters are also used to start controlled burns during ideal conditions. Because they can fly slowly or hover, they are used much like a giant flying drip torch over a large area.

Controlled burns for fire prevention during the off season happen all the time in Oregon. I talked about how they're done on a fire or a complex, and a little about slash and burn, but not the process of bidding and then doing a controlled burn on government property. Controlled burns are a part of most, if not all, timber contracts. The USFS, Bureau of Land Management or Oregon Department of Forestry will first send an official to inspect and mark out the land they will put up to log. The inspector will walk the land and nail or staple signage to the trees to mark the area that can be logged. They will create a map of the area and start to write up the contract on the approximate market value of the trees. This depends on the type, age and density of the trees. There are many reasons to harvest timber: for the money, to thin a section of the land to prevent wildfires, to harvest usable timber in an area that has already been burned over and more. Sometimes only the bark or branches of a tree will burn, and sometimes there is salvageable wood in the burnt trees.

Ralph Davis owned a contracting company for over twenty-five years. Companies like his would bid for these timber contracts, and the contracts were very complicated. For instance, if the land belonged to the USFS, the

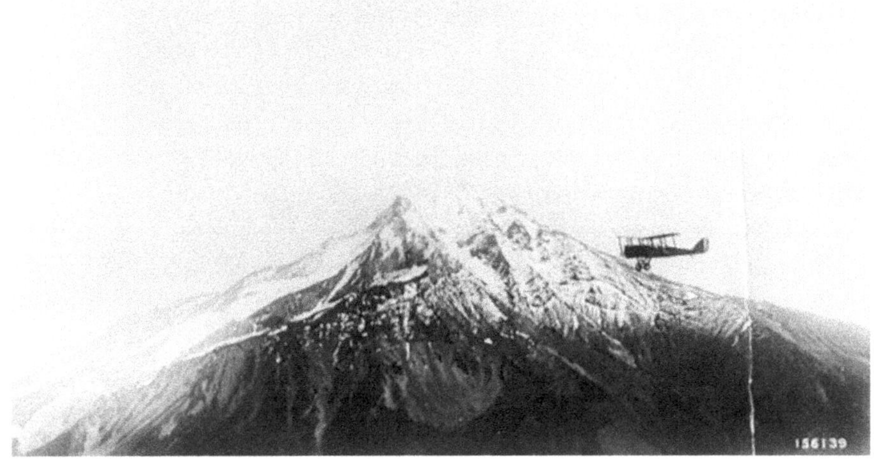

Fire patrol airplane circling Mount Jefferson in the Cascade Range, Willamette National Forest, 1920. *Photograph by U.S. Army Air Service, courtesy of the Gerald W. Williams Collection at Oregon State University.*

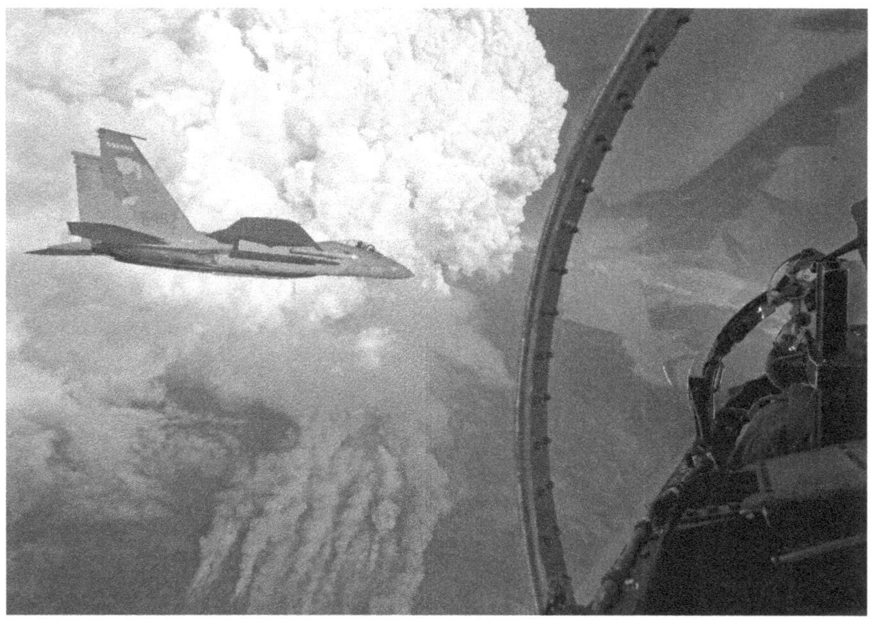

F-15 Eagles from Oregon Air National Guard 173rd Fighter Wing fly over the wildland fires in Southern Oregon following a routine training mission. *Photograph by Jim "Hazy" Hazeltine, courtesy of Wikimedia Commons.*

money made would go to the U.S. Treasury, and a portion of it would end up coming back to various divisions inside the USFS in its annual budget. The timber companies would bid for the contract. Once a company is chosen and the contract is signed, it sends out a team of loggers to cut down the trees, but this is only half of the process. The timber company usually has a total of five years to rehab the land and replant a new generation of trees. To do this, the timber company would hire a contract team like Davis's crew, and they would come burn the slash. During the first phase of the process, a tree is cut down and the loggers cut off the top and branches. Anything smaller than about twelve inches in diameter is cut off and discarded because it cannot be turned into useful timber. Because of this, a timber company clear-cut is filled with stumps, discarded wood, vine maples, scrub brush and other vegetation. All of this needs to be burned and cleared.

The slash and burn crew preps the land by making piles of wood, cutting a fire line around the area and performing other preventative actions determined by the type of fuel, weather, time of year, et cetera. If the property is next to power lines or structures of any type, the crew

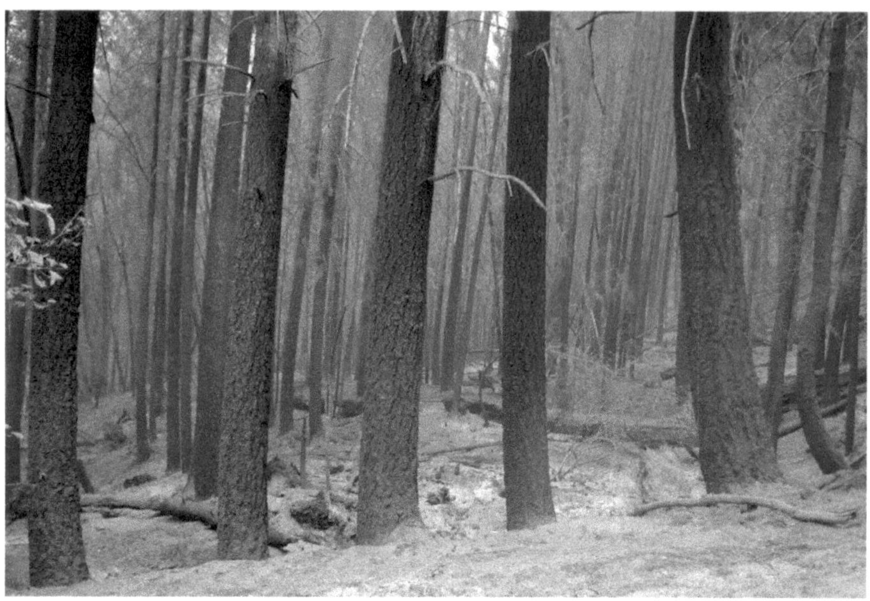

Trees in the black can have their bases burned and may have the potential of falling over due to wind or other factors. Because of this, they become a hazard to wildland firefighters and the public. These snags can be salvaged for wood, so the timber is sometimes harvested. When this is the case, the job will be put up to bid. The job can be bid on by multiple contractor crews, and sometimes different phases of the job will be done by different crews. For example, one may do the slash and burn while another replants trees after all the land is cleared. *Courtesy of the U.S. Forest Service.*

may set up a hose lay and sprinkler system all around the perimeter. Ralph Davis's crew, on the fire pictured opposite, had to set up a hose lay with a sprinkler system all the way around the fire because it was so close to structures.

Most of the time, burning means staggering five or six men or women and having them walk across the highest part of the land. The person on top walks first, the next person gives them five or ten feet and starts walking, followed by the person below him or her and so forth. The person on top has to go first. If the person at the bottom went first, the fire may burn too fast, trapping the people on top. Crews don't want to set the fire at the lowest point, because fire always climbs terrain. There are problems with lighting fires from the top, too. If the land is too steep and unsteady, wood that is on fire may roll downhill and get in front of one of the wildland firefighters burning with a drip torch. The rolling fire can hit the firefighter or start a fire in a place where the crew has no

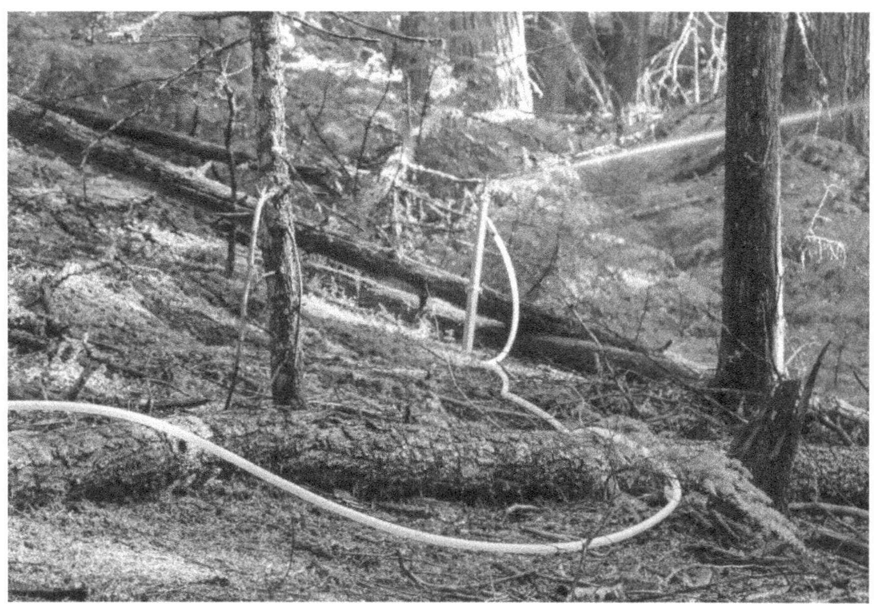

Sprinkler kits come in a big box with all the parts and pieces a wildland firefighter will need to set up four sprinkler heads. The kits come with a toy house, a hammer, nails, zip ties, wood blocks and much more. *Courtesy of the U.S. Forest Service.*

control over it, so an experienced crew boss will decide where, when and if they will start the burn operation. Some controlled burn missions are refused. If it is determined to be too dangerous to burn on foot, there are a few other options; one of them is shooting magnesium rounds through a flare gun.

Another way of burning difficult terrain is by helicopter. This means that the contract crew will have to hire a contract helicopter crew, but the job will get done. The job of burning a clear-cut is usually done within a year of cutting the trees. Any longer than a year and more vegetation will grow and it becomes more difficult, but after the land is burned, the tree planters usually have five years from when the original trees were cut. The whole process takes up to five years, and a lot of the money for each stage in the process is held by the USFS or Oregon Department of Forestry until the job is done.

Helicopters aren't used for controlled burns very much these days. Today, they are used for putting out fire, dropping retardants or increasing the relative humidity in an area so the fire won't spread. When a fire makes a run or burns in very difficult terrain, or if the leadership team

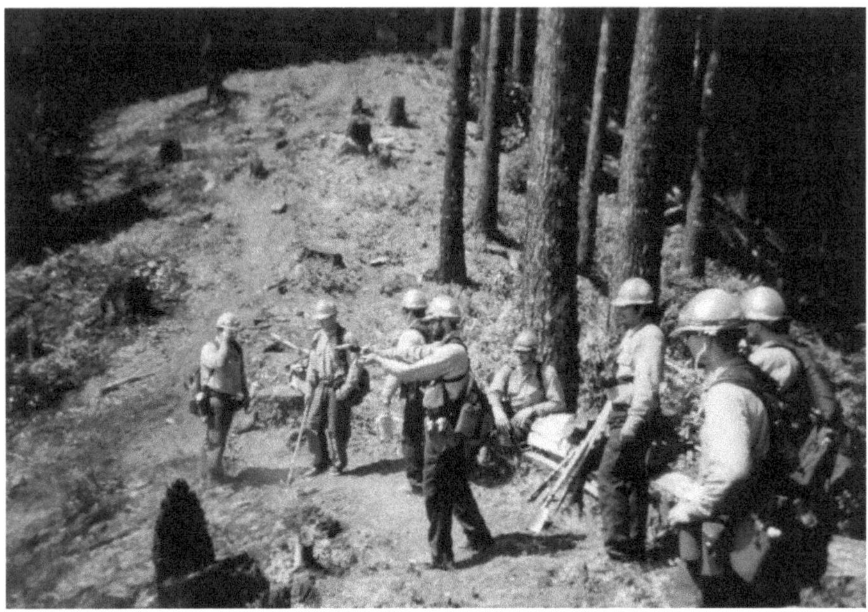

Zigzag Hotshot crew blackening the line. They had used drip torches to get what they could at the top and then used a magnesium flare gun to burn the bottom of a ravine, 2008. *Photograph by Tyee Burwell.*

believes the use of helicopters can help contain the fire, helicopters will be used. Helicopters can haul large buckets or be fitted with tanks and drop the water from the air. Specialized helicopters like the Erickson AirCrane have front-mounted foam cannons. Helicopters with buckets fill them in various ways. They can dip in a pond, reservoir or some other body of water, or the buckets can be filled while on the ground by hoses pumping from any of these bodies of water or a water tender (a truck with a large tank of water), pumpkins (portable tanks) or a number of other sources.

Helicopters also carry crews to remote areas and either land there or let the crews rappel down to the ground. Once the personnel have rappelled, equipment is lowered down to them. While this is a great way to get a team into a remote area to fight fire, the team must have enough water, food and equipment for the duration of the time they will be deployed. They also must hump all the equipment they bring into their extraction zone. It takes an extraordinary amount of endurance and focus to be on a fire helitack crew.

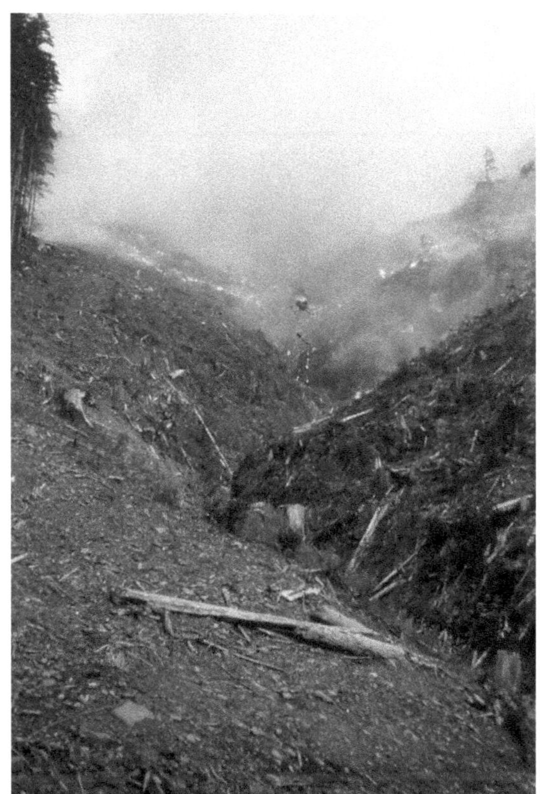

Right: Controlled burn by helitorch, 1978. This is essentially dripping napalm from a helicopter onto a unit of land after a timber harvest. The U.S. Forest Service no longer does this. *Courtesy of Margaret Beilharz.*

Below: The helitorch was used to burn steeper and more remote units. There is a much higher probability of injury when wildland firefighters burn with drip torches on steep landscapes, as fire will travel quickly up a steep slope. This photograph was taken in 1978. *Courtesy of Margaret Beilharz.*

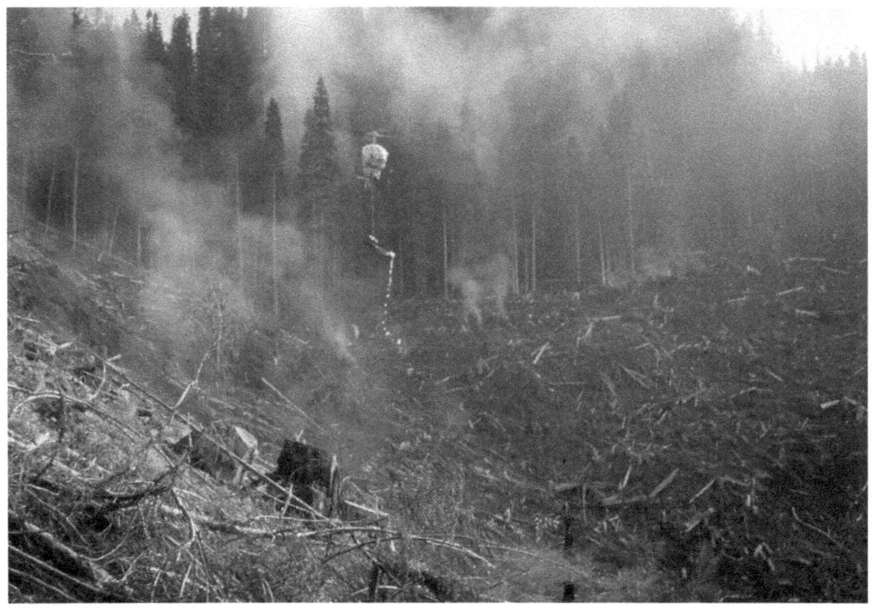

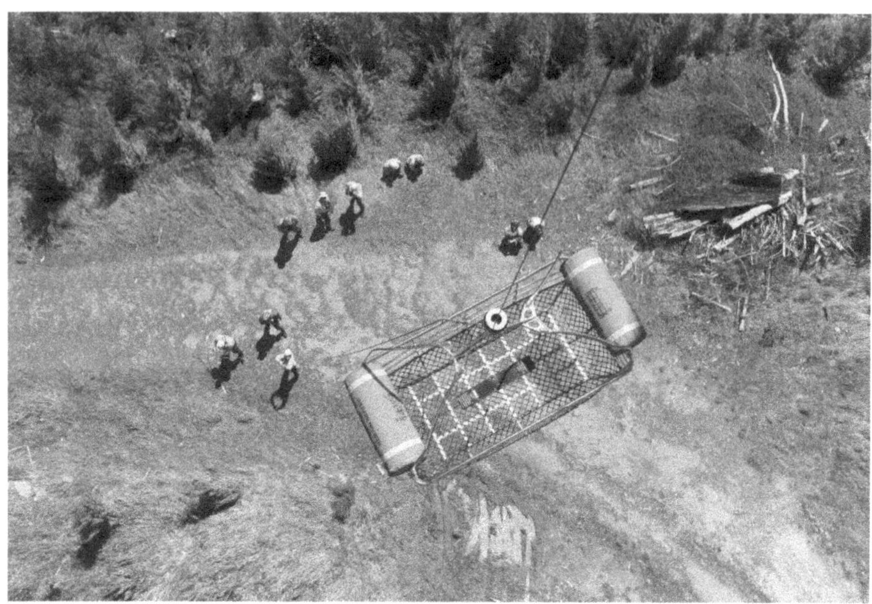

Helicopter training to extract injured personnel. *Courtesy of Wikimedia Commons.*

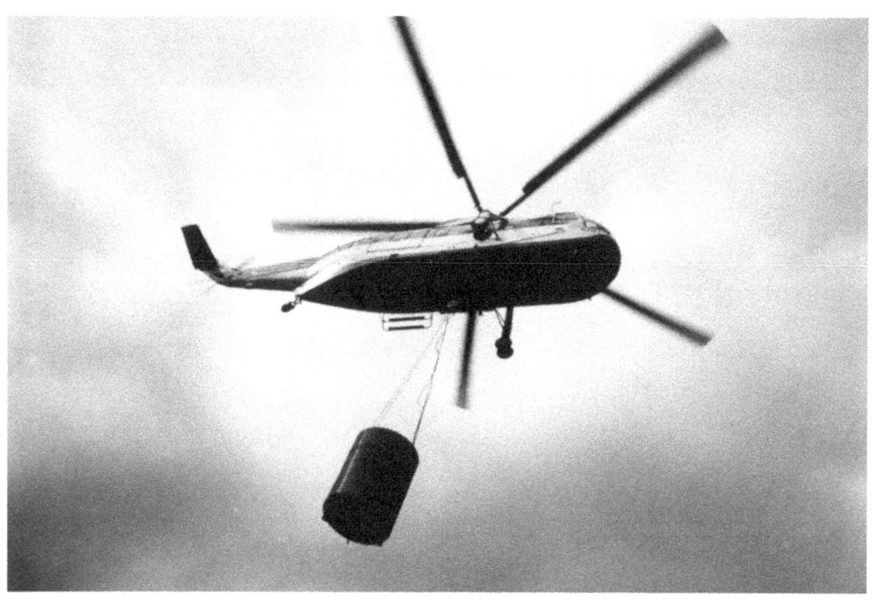

Here is a B-17 making a fire retardant drop on the Tool Box Fire, Silver Lake District, Fremont National Forest, 1966. *Courtesy of the Gerald W. Williams Collection at Oregon State University.*

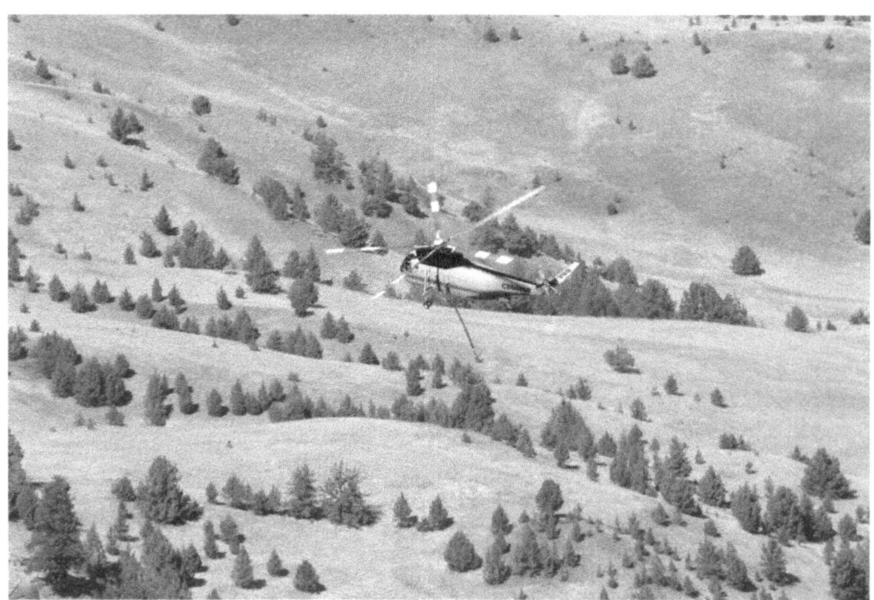

Sikorsky S-61L helicopter flying to drop water on the Lost Hubcap Fire near Monument, Oregon, 2014. *Courtesy of the U.S. Forest Service.*

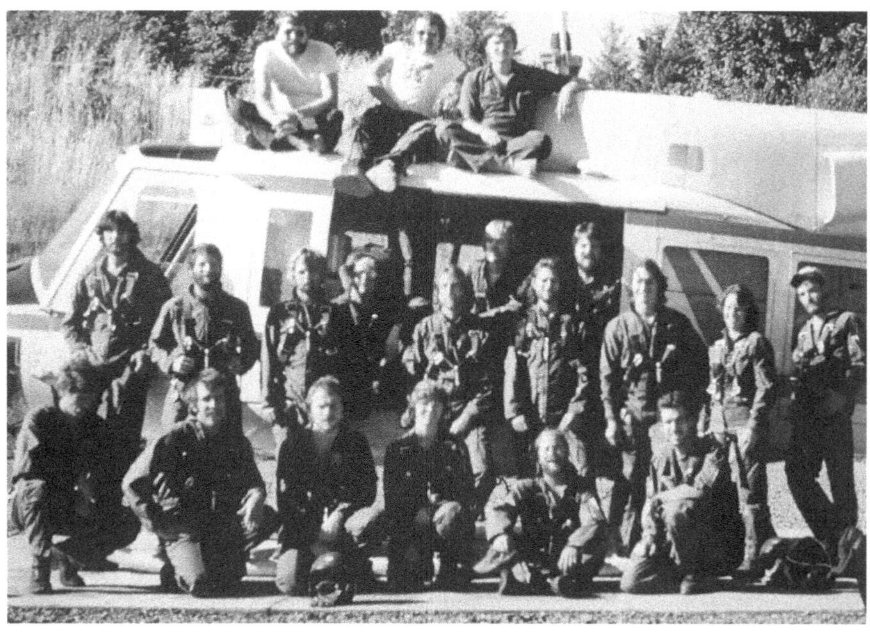

The Rigdon Rappellers, 1981. *Courtesy of Valerie Rapp.*

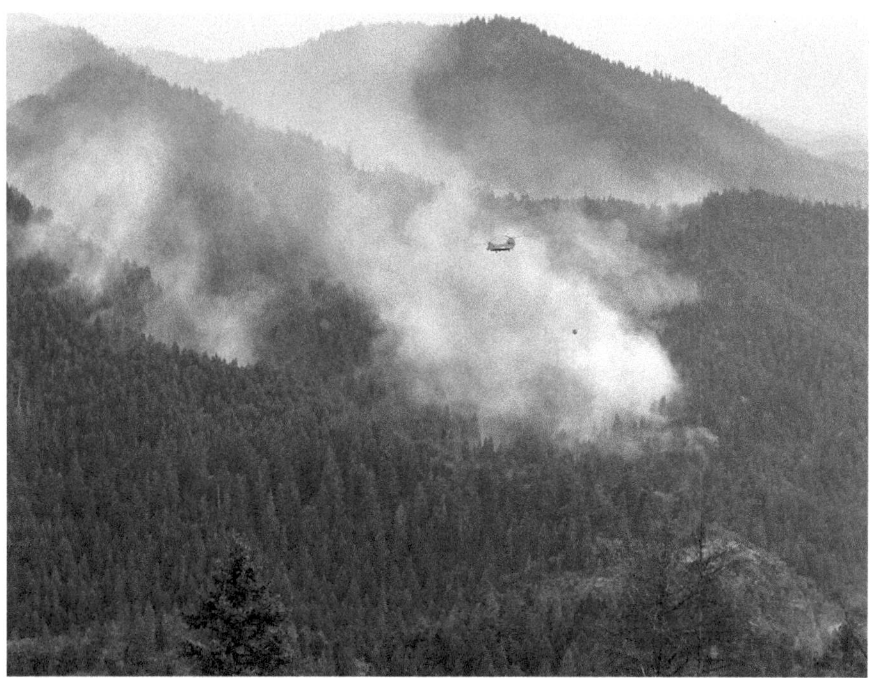

A Chinook helicopter coming in on a bucket drop on the Taylor Creek Fire in Southern Oregon, 2018. The Taylor Creek and Klondike Fires grew into each other despite attempts to put them out separately. Combined, the two fires burned more than 225,000 acres. *Photograph by the author.*

The helicopter used really depends on the fire, weather, terrain and mission, but the most commonly used helicopters are the Bell 204/5, Bell 212 (the Twin Huey), the Sikorsky S-61, the Sikorsky S-64 SkyCrane and the CH-47 tandem rotor Chinook.

Fixed-wing aircraft, or planes, have been used to fight fires for about one hundred years. While the USFS says that they are critical tools, the agency also says that planes won't put out the fire by themselves. That said, planes are incredibly useful and have a number of important roles. We've covered that planes are used to spot fires and smoke columns in wilderness areas. They transport people and equipment. Smokejumpers jump out of planes to a place where they could do the most good, and planes put a lot of wet stuff on the hot stuff. The Evergreen 747 Supertanker can hold up to 19,600 gallons of water or retardant. The smallest fixed-wing in use is the AT-802F, or air tractor, and it can hold up to 807 gallons. There are a

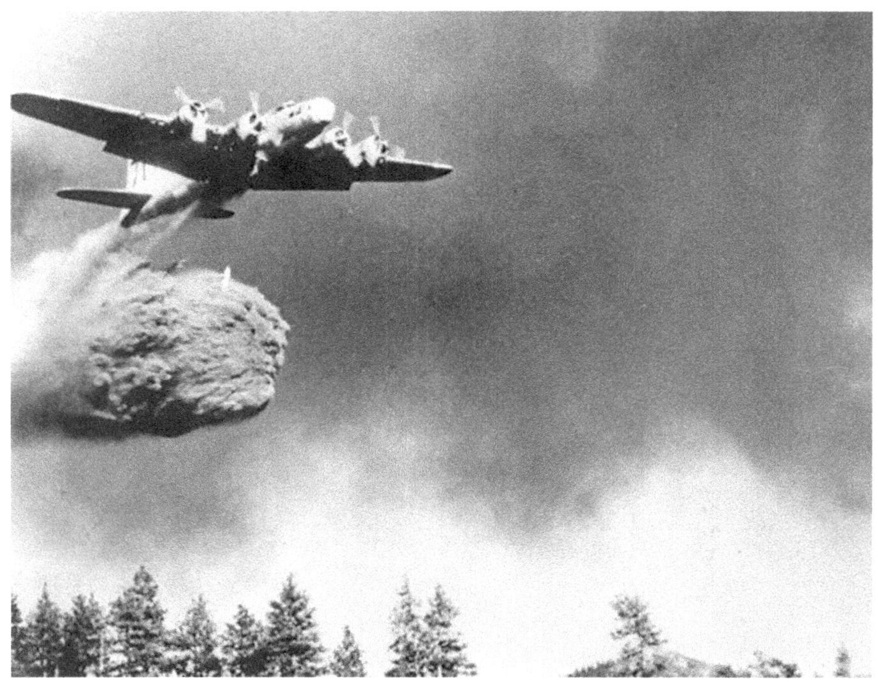

In this photograph, a B-17 makes a fire retardant drop on the Tool Box Fire, Silver Lake District, Fremont National Forest in Lake County, 1966. *Courtesy of the Gerald W. Williams Collection at Oregon State University.*

number of options in between, and every one of them is a beautiful sight for a wildland firefighter on the ground.

Once the fires are out, there is still a lot of work to do in the Oregon forests. The USFS and Oregon Department of Forestry rehabilitate trails, mitigate fuels around a town, mark timber to be sold or slash burn after a fire. These organizations are having a difficult time lately finding money in their budgets to do these jobs in the off-season, especially since the cost of wildland fires has increased. This pattern of hotter and longer wildfire seasons will only increase. The Oregon Climate Change Research Institute says:

> *Over the last several decades, warmer and drier conditions during the summer months have contributed to an increase in fuel aridity and enabled more frequent large fires, an increase in the total area burned, and a longer fire season across the Western US. The decline in mountain snowpack and earlier spring melt contributes to the fire season. These seasonal droughts*

> *will occur under warmer conditions in the future, stressing trees directly by exacerbating water limitations.*[19]

The institute goes on to say that the fire season in the Pacific Northwest has increased in length every decade for the last forty years. In the 1970s, the season was, on average, 23 days long. In the 1980s, the average fire season lasted for 43 days. An average fire season lasted for 84 days in the 1990s, and in the 2000s, the average fire season was 116 days.[20]

In 2017, the USFS spent $2 billion on fighting wildfires across the country, and according to Donald Trump's budget, it was to receive only $4.7 billion for all USFS programs. This is $787 million less than it received the previous year.[21] The year 2017 was the most expensive year of firefighting since 2015 ($1.7 billion), and the wildland fires, especially in Oregon, are only getting worse.

While the CCC is no longer is a federal organization, there are states that have similar programs. The California Conservation Corps gives at-risk youth opportunities to be trained to do jobs on a wildland fire. Oregon has a few organizations that train youth in forestry jobs and give them the opportunity to find careers in the field while learning leadership skills and self-confidence. The Northwest Youth Corps, Heart of Oregon and the Central Oregon Youth Conservation Corps are a few of these organizations, and they mostly rely on donations and grants to fund programs.

Whether it's fighting wildfire, rehabilitating after a fire or building trails and reforesting after a fire, wildland firefighting agencies are always trying to adapt and improve. Today, these agencies are looking at thermal imaging technologies. On the Rebel Complex, a 2017 fire in the Willamette National Forest, the USFS was able to use thermal imaging technology to see through the smoke while flying over the fire. While it didn't give the leadership a crystal-clear image, it did let them know the head of the fire had shifted, and because of that, the evacuation notice was not upgraded and people were able to stay in their homes.

Fire surveillance drones also present a big possibility in the future. Currently, if a drone is in the air, it grounds all aircraft. It's a safety issue, but if an incident commander could be able to put a thermal imager on a drone and get real-time information on fire behavior, this could save lives.

Today, the glue that binds the fire shelters carried by every wildland firefighter melts at five hundred degrees Fahrenheit, and the blanket itself is made of an outside layer of high temperature–resistant silica cloth with an inside layer of fiberglass scrim cloth. Imagine a fire shelter made of a

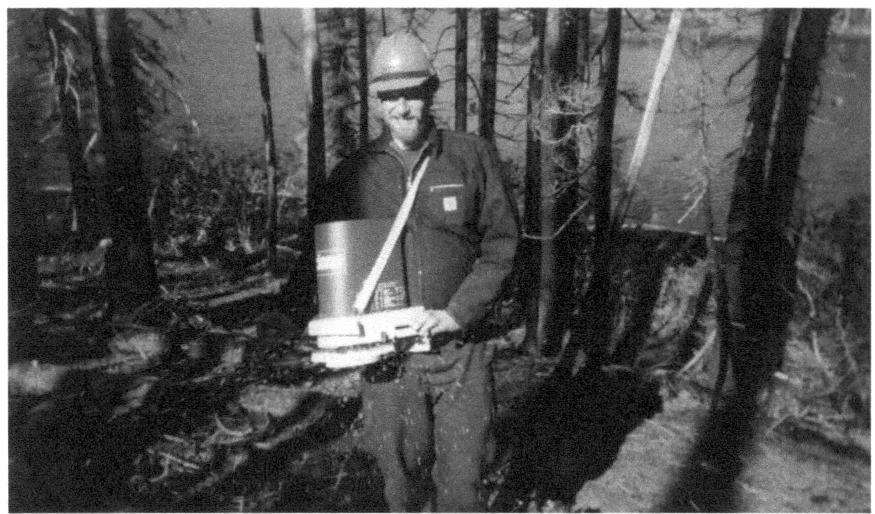

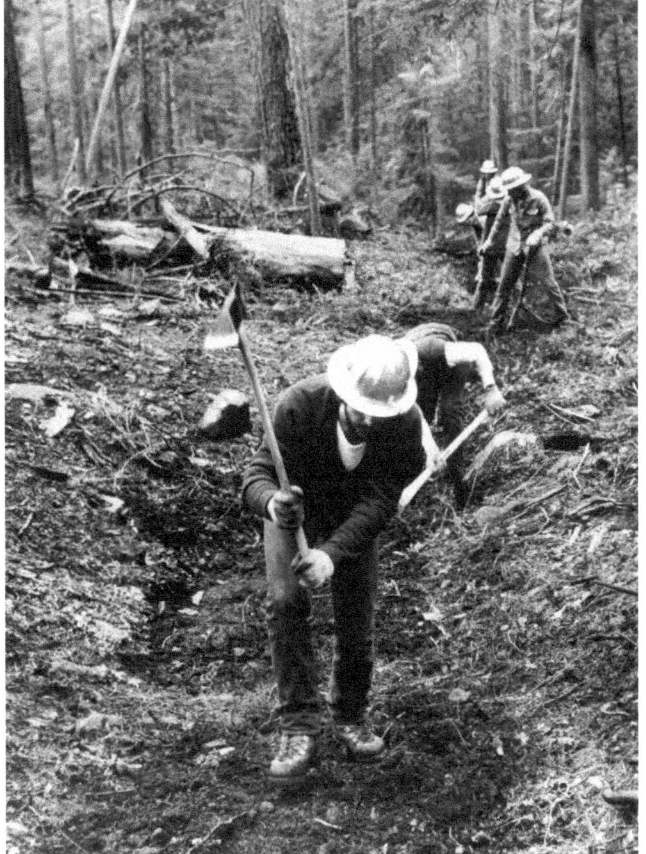

Above: Tyee Burwell with his U.S. Forest Service crew planting grass seeds at an area blackened by the Ollalie Lake Fire, 2003. Their mission was reseeding and soil stabilization. *Courtesy of Tyee Burwell.*

Left: Young Adult Conservation Corps corpsman builds a fire trail, Blue River District, Willamette National Forest, 1977. *Photograph by Sam Frear, courtesy of the Gerald W. Williams Collection at Oregon State University.*

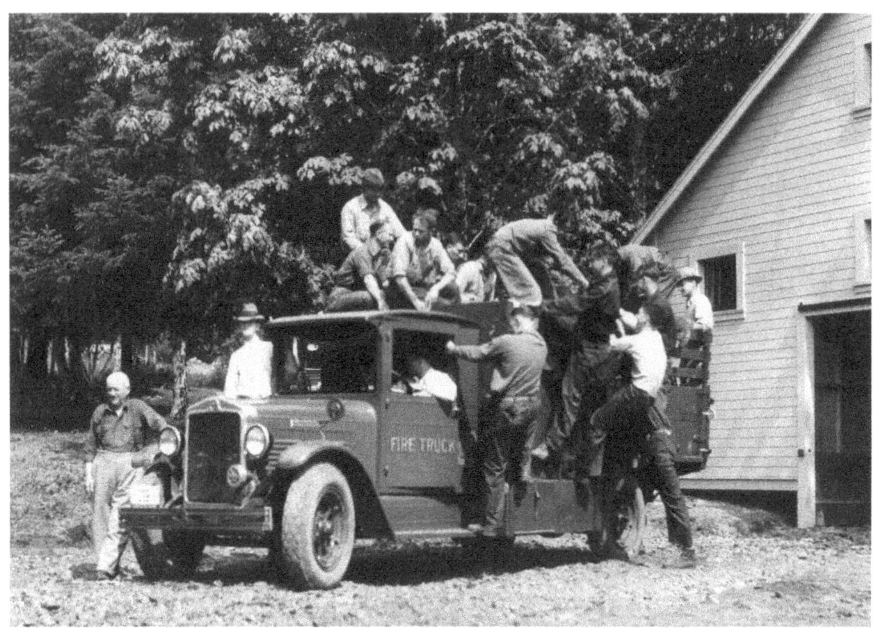

CCC crew jumping on their engine and heading to a wildland fire, 1933. *Photograph by K.D. Swan, courtesy of the Gerald W. Williams Collection at Oregon State University.*

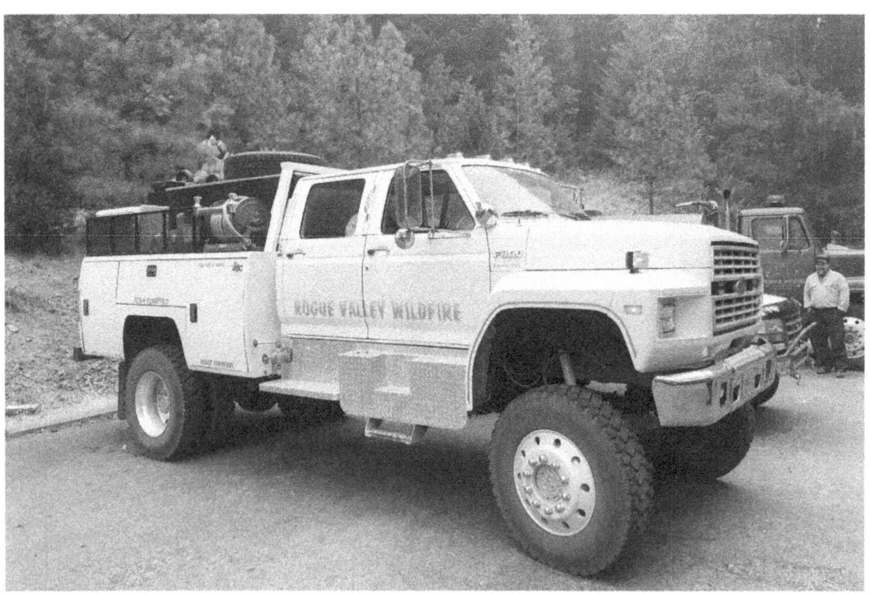

A Type V contractor wildland firefighting engine from Rogue Valley. An ICS Type V engine has a tank capacity ranging from 400 to 750 gallons. This was on the North Umpqua Fire in Southern Oregon, 2017. *Photograph by the author.*

ceramic-based material that would reflect radiant and convective heat up to three thousand degrees Fahrenheit.

In addition to technological advances, there will also be advances in communication, leadership roles and skills, fire suppression and retardants and much more, but no matter how far we go, we will always need the brave men and women in their greens and yellows on the ground with boots on their feet and tools in their hands.

Chapter 7

THE LARGEST WILDLAND FIRES IN OREGON

The Great Fire, 1845

Wildland fires in Oregon happen every year. There's no way around it. The majority of the state is flammable, and that will always be a part of our history and our future. One of the worst and earliest fires in United States history happened here in Oregon. This fire was known to generations of people simply as the Great Fire. In 1845, it burned 1.5 million acres, half of Lincoln County and half of Tillamook County, from the Willamette Valley to the coast. We may have never known the origin of this fire if the editor of the *Sheridan Sun* hadn't traveled to the coast and stood in the middle of the burned area. He said there was nothing but old-growth snags as far as the eye could see in every direction. It was the largest old-growth forest to burn in a single fire in the whole of the United States. It still is.[22]

While on his journey, the newspaper editor met a commercial fisherman named Peter Belleque.[23] Belleque's father was French, his mother was Native American, and he was born in 1836. The Great Fire started when he was nine, and he knew why. The story claimed that an English ship entered the Columbia River, and on this ship was a cook named Johnson. Johnson hated being a cook on an English ship, so at the first chance he had to get off the ship and try to make a life, he did. Johnson deserted his duties and his captain and fled down the Willamette Valley to what we now today Champoeg. He staked his claim and started to develop his property by cutting and clearing it. One day, he thought the conditions were right to burn the brush. He was wrong.

The Willamette Valley is covered in tall grass, shrubs, forest and, at that time, old-growth virgin timber that was as old as six hundred years. He didn't count on the wind direction changing, which it did, setting off a fire so intense that no one has seen the likes since.

Belleque said he remembered the hot wind and smoke being so dense that settlers had to eat their midday meal by candlelight. Belleque had talked to other people who lived in that area at the time of the Great Fire. He was told tales of animals found boiled in the water they jumped into in order to escape from the flames. The fire displaced tribes of Native Americans, and many people thought the end of the world had finally come until, weeks later, heavy rains put out the fire.

The Silverton Fire of 1865

On any list of the largest fires in the United States, you will find Oregon a number of times. The people of Oregon only had to wait twenty years for another great Oregon fire. There was a decent-sized fire in 1853 called the Nestucca Fire that burned 320,000 acres, but the Silverton Fire dwarfs it in comparison. The Silverton Fire also started in the Willamette Valley and burned 988,000 acres. This fire burned through so dense of a forest and fuel that it left ash ten inches deep.

The Columbia Fire or Yacolt Burn, 1902

This series of fires grew from lighting one pile of brush on a slash and burn to clear the land for development, and it grew with other fires on the Oregon and Washington border to a total size of 640,000 acres. The fire completely destroyed the town of Palmer. The fires ended up taking sixty-five lives and destroying an unknown amount of structures. After falling from the sky, an inch and a half of ash covered Portland. Street lamps in Seattle came on at noon. The fire came close enough to the town of Yacolt that it blistered the paint on fifteen buildings. It also should be noted that Yacolt is a native word meaning "place abounding in evil spirits."[24]

The Big Burn, The Great Fire, The Big Blowup or Devil's Broom Fire, 1910

The Great Fire didn't happen in Oregon, but any book about wildland fire should at least mention the Big Burn. This is the fire in which Ed Pulaski held his men inside an abandoned mineshaft at gunpoint to keep them from giving in to fear and running into the fire to be burned alive. His crew were so afraid of the sounds from the hurricane-force winds and the extreme heat they were experiencing that they believed that they needed to escape, but Pulaski pulled his pistol and leveled it at them to keep them in the cave. All but five of the forty men lived. This fire burned three million acres across Montana, Idaho and Eastern Washington. Twenty-eight men, or the "Lost Crew," were burned over by flames and died on Setzer Creek outside of the town of Avery, where people were still trying to evacuate while the flames started burning it down. Drought, high temperatures, an early

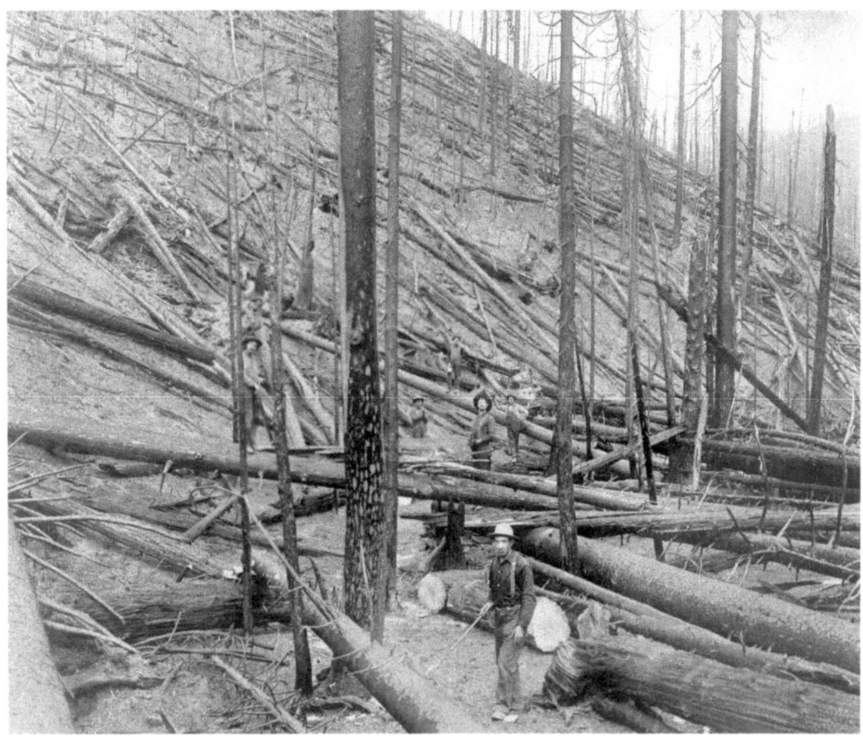

Result of the wildfire "hurricane" in a heavy stand of Idaho white pine on Little North Fork of St. Joe River, Coeur d'Alene, Idaho, 1910. *Courtesy of the National Photo Company Collection of the Library of Congress.*

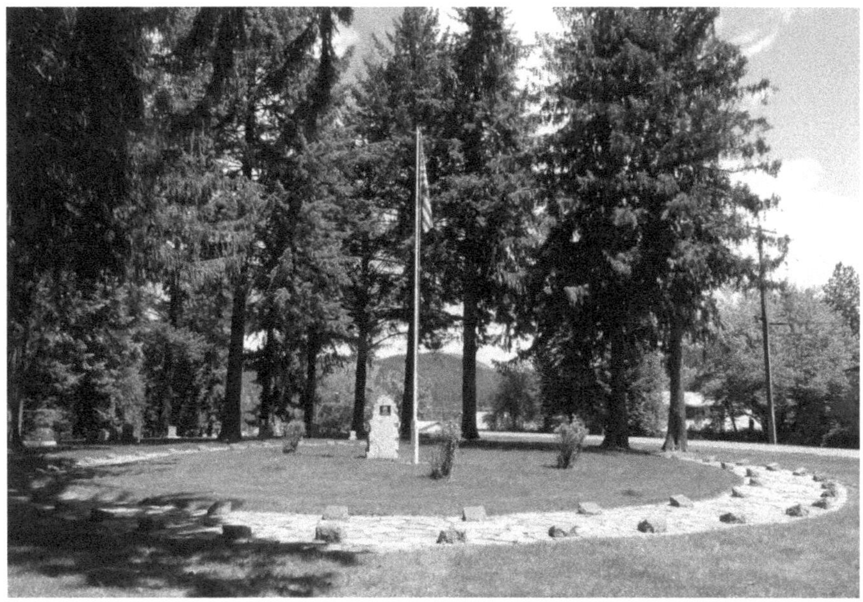

The memorial to firefighters who died in the Great Fire of 1910, located in Woodlawn Cemetery, St. Maries, Idaho, June 2015. *Photograph by Ian Poellet, courtesy of the Creative Commons Attribution–Share Alike International License.*

start to summer, the right type of fuel and unusually strong winds all made conditions perfect for this fire to really go. It only took two days for it to burn 4,700 square miles.

The best guess about the origin of the fire was the newly constructed Chicago, Milwaukee & Puget Sound Railway that ran through the densely forested Bitterroot Mountains. The coal-powered locomotives frequently shot red-hot cinders into the forests. The spot fires from the trains, along with previous lightning storms, caused up to three thousand fires at once. The second day, most of them combined and grew so fast that entire towns were in danger. Flames shot hundreds of feet in the air "fanned by tornadic wind so violent that the flames flattened out ahead, swooping to earth in great darting curves, truly a veritable red demon from hell."[25] The winds grew so strong that they sucked flaming trees from the ground and threw them. The fire burned so intensely that it was seen in upstate New York and Colorado. The smoke filled so much of the sky that boats off the coast of Washington and Oregon could not use the stars to navigate. Soot from this hell fire fell on the ice in Greenland. People across the country did their best to make peace with their god, because the end of the world had surely come.

But on August 22, just two days after it started, the winds calmed, and it rained and snowed in higher elevations. In the end, the fire destroyed eight billion board feet of lumber and twenty million acres and took eighty-six lives. A memorial still stands in Woodlawn Cemetery in St. Maries, Idaho, for the firefighters who died in the Great Fire in 1910.

THE TILLAMOOK BURN, 1933, 1939, 1945, 1951

The Tillamook Burn is the collective name for a series of fires that struck every four years from 1933 to 1951. The first fire was the largest, and it was started on August 14, 1933, by loggers in steep terrain above the North Fork of Gales Creek, just west of Forest Grove. The loggers said this fire started by the friction of dragging a Douglas fir across a downed tree in the middle of a grove of very dry fuel on a 104-degree day. In less than sixty minutes, despite a valiant effort to put it out, the fire burned through 60 acres of slash. Due to a lack of roads through the forest on the coast, the fire took off. In ten days, the fire had blown up to 40,000 acres, but on the night of August 24, the fire ballooned to 240,000 acres in less than one day. The firefighters, which now numbered over one thousand, could do nothing but watch.[26]

According to historian Doug Kenck-Crispin, "In the Burn, after the 24th, ships 500 miles at sea reported flaming debris landing on their decks."[27] The trees that didn't completely burn stood and dried a little more every year along with the forest untouched by flame. Then, six years later, it started again, and 209,000 acres burned. This included 19,000 acres that hadn't burned in the first fire. This same thing happened again six years later. In 1945, two wildland fires burned over 182,000 acres, and this time, almost all of it had been untouched by the two previous fires. Six years later, it happened again. This time it burned 32,700 acres.

The Tillamook Burn seemed to be a cycle as certain as the seasons, but Oregon was able to break the cycle. Landowners, still feeling the effects of the Great Depression and not wanting to pay property tax on useless burnt land, forfeited their land to the counties. The counties transferred the lands to the state, and when the timber was harvested, the state received a part of the profit. With this money, the Oregon Department of Forestry began programs for fire protection and reforestation.

Between 1949 and 1972, when the brush disposal (BD) funds were flowing, the Oregon Department of Forestry planted more than 72 million

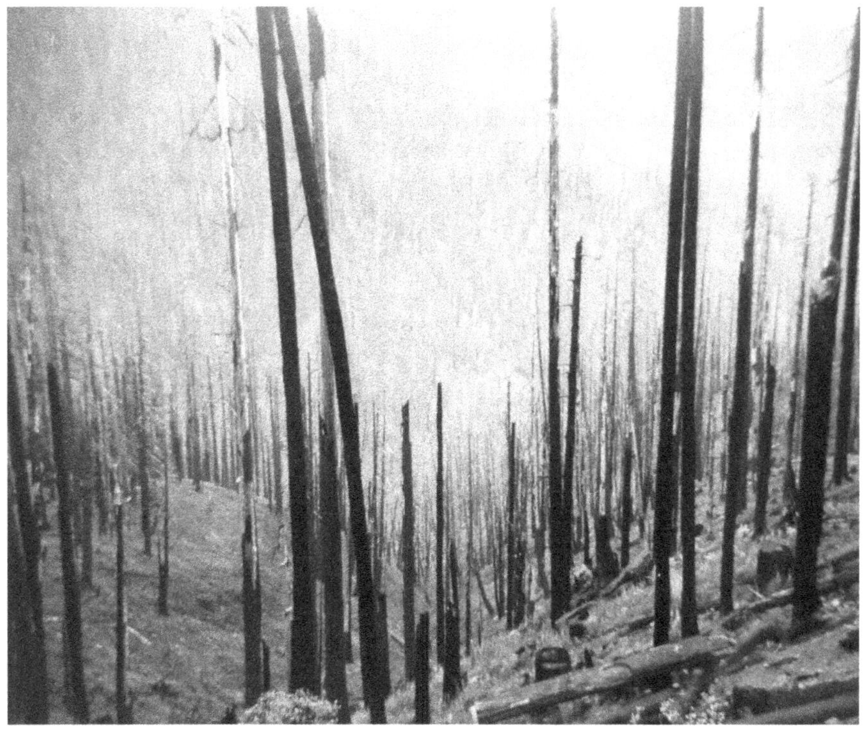

Tillamook Burn, photograph by Russell Lee, 1941. From a black-and-white negative from the Farm Security Administration Office of War Information. *Courtesy of Library of Congress.*

seedlings in the burned area, and they dropped over a billion Douglas fir seeds from helicopters. Schoolchildren from all over the state were driven there on school buses to plant almost one million seedlings. In this way, the reforestation effort changed the Tillamook Burn into the Tillamook State Forest.

The Bandon Fire, 1936

To understand how this fire wreaked so much havoc and completely destroyed the business district and docks of Bandon, you need to know the origin of the town. In 1873, an Irish immigrant named George Bennett (1827–1900) and a handful of other new Oregonians settled an area at the mouth of the Coquille River. Bennett convinced those already in the area to combine their two settlements and name the town after Bandon, a town in in Cork County, Ireland.

To further make his town more like the town of his motherland, Bennett planted an ornamental shrub common in Ireland called the Irish hedge. Thirty-six years after Bennett had died, this invasive species of gorse filled most spaces between the buildings of the town. On September 26, 1936, a forest fire burned a few miles out of town. It was close enough that people could smell and see the smoke, but those in Bandon weren't overly worried until the wind changed direction. In only a couple of hours, the small pieces of burning wood flying with the wind landed in the gorse. It ignited immediately. Bandon resident D.H. Woomer was quoted in the *Coos Bay Times*: "That Irish hedge was the worst thing—when the fire hit it right across from my house, the flames shot up high into the air. It was just as though there had been gasoline poured on the fire. And water was just no good against it—wouldn't touch it! The stuff seemed just full of oil."[28]

A firestorm flew through town, destroying all of the business district plus one hundred or more homes. Most of the people in the town escaped the flames, but not all; ten lives were lost in the fire.

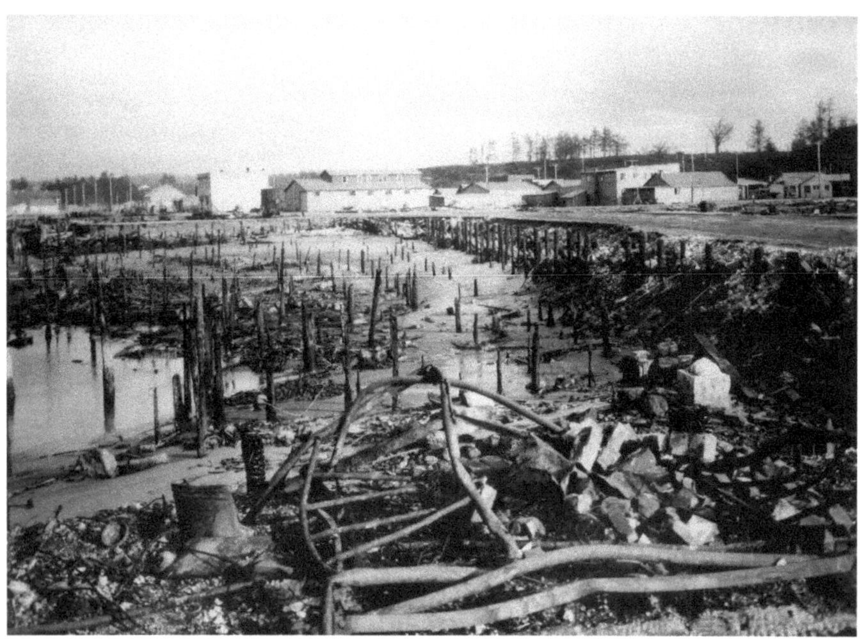

The docks and business district of Bandon burned within hours. *Courtesy of Bandon Historical Museum.*

The Ashwood-Donnybrook Fire, 1996

The Ashwood-Donnybrook Fire is interesting because it not only burned 118,000 acres but also started on unprotected lands. Earlier in the book, I noted that some lands in Oregon aren't protected by the USFS, the Oregon Department of Forestry or any other fire agency because the residents in the area had elected to not pay extra money in their taxes for the protection. Fourteen percent of the 41 million acres of Oregon are unprotected lands, and the official policy is to let owners decide to fight it or let it burn unless it poses a danger to structures or timberland. The Ashwood-Donnybrook fire raged through the sage of Eastern Oregon, and by the time it left the unprotected land and crept into state-protected territory, the head of the fire was ten miles wide.

When a fire burns through the plain, it is actually beneficial. It burns away debris, kills pests and adds nitrogen to the soil, so landowners let the fire burn. But in this case, letting the fire burn cost hundreds of thousands—if not millions—of dollars. This and other fires like it caused legislation to be proposed in the late 1990s and again in 2003 that declared fires on unprotected lands to be a nuisance. It would allow state governments to bill landowners for the cost of fires; this didn't fly. The compromise was to allow landowners the choice to authorize nonprofits called Rangeland Fire Protection Associations to fight these fires. These nonprofits receive equipment donated by a number of private and public agencies along with training from the state. These volunteer nonprofits lack the resources of other agencies, but they can get to the fires faster and put them out before they get too big.[29]

Simnasho Fire at the Warm Springs Reservation, 1996

Warm Springs Reservation has fire every year. I personally have fought fires there for the last four years in a row. The fires burn through the brush and sparse ponderosa pines, so it moves fast, but the wind also completely changes direction every afternoon, which keeps the head of the fire moving from one place to another. On the reservations, the Bureau of Indian Affairs (BIA) controls the incidents with the help of other agencies. In 1996, a fire caused by engine exhaust started about ten miles north of Warm Springs in an area called Simnasho. This fire burned through grass, brush and timber until 118,000 acres were black. This happened despite having 1,646 total personnel, 52 line crews, 113 engine crews and 7 helicopters working on the fire.[30]

JACKSON FIRE, 2000

The Jackson Fire was significant because it burned over 100,000 acres, but it grew so much that it threatened the Snake River Correctional Institution. The state had to figure out how to evacuate three thousand prisoners and where they would take them. To try to solve the problem, Governor John Kitzhaber called for a conflagration. The Emergency Conflagration Act was first passed as an Oregon Revised Statute in 1940. When a governor or other authorized official deems a fire an emergency that threatens lives or property, he or she may declare it a conflagration. This means that the governor can assign and make available any county, city or district resource under "the command of an officer designated by the Governor for the purpose of, any part of the fire-fighting forces and equipment of any fire-fighting organization in this state other than an organization that possesses only one self-propelled pumping unit."[31] In a nutshell, this means that the governor can direct any firefighting unit or organization in the state to fight a particular fire; this includes structural fire departments. Not only that, but the government will pick up the tab on overtime and equipment. Once the fire was declared a conflagration, the Baker/Union, Umatilla, Wasco and Snake River/Malheur County structural agencies all came to help fight the fire. They did lose one manufactured home, a satellite building, an RV and a horse trailer, but they saved 150 houses and the correctional institute.

THE LAKEVIEW COMPLEX, 2001

A thunderstorm filled the night sky on August 9, 2001. Hundreds of lightning strikes ended up starting a number of fires around Juniper Mountain in Malheur County. By the time the storm was over, at least four large fires were raging. These fires—the Horsehead Fire, Mustang Fire, South Warner Fire and Johnson Creek Fire—all grew into the Lakeview Complex that went on to burn more than 127,000 acres in only seven days. This is an incredible rate of growth, but wildland firefighters were able to get this one 100 percent contained.

The Biscuit Fire, 2002

The summer of 2002 was one of the state's hottest on record. During this heat wave, from July 12 to the July 15, a storm cell covered much of Oregon, and in those four days, there were over 1,200 lightning strikes, which led to 375 different fires. On July 13, a large thunderstorm rolled in above the Kalmiopsis region just inland from the Wild River Coast of Southwest Oregon and Northwest California. Lightning struck the Klamath-Siskiyou Mountains again and again. The first fire was spotted at 1537 hours and burned less than an acre of land, but in two weeks, it had grown and absorbed other fires and had burned into California. In another week, over seven thousand personnel were fighting this fire, with some firefighters coming from as far away as Australia and New Zealand.

The fire grew into the largest Oregon had seen in almost one hundred years. To make matters worse, 2002 was a horrible year for fires, and most of the wildland firefighters had been on other fires before being assigned to the Biscuit Fire. The flames easily jumped the Illinois River and threatened commercial forest and structures.

Ridges, roads and fire lines would determine the size and shape of the fire perimeter. The incident command decided to blacken as much as 100,000 acres of this 500,000-acre fire as a suppression strategy. Some of these controlled burns got out of control and needed suppression strategies of their own.[32]

Amazingly, no one died on this fire, one of the largest the state has seen. In fact, half of the fire "burned cool," meaning that much of the vegetation in the black was unburnt. The Biscuit Fire burned from the day it started in July and kept burning until the USFS officially called it out on New Year's Eve. It did burn a few cabins on mining claims and some private land and put fifteen thousand people on evacuation notice.

The USFS went in and logged the burnt areas and salvaged up to sixty-seven million board feet of fire-killed timber. They only logged 1 percent of the total burned area. A controversy about whether the burned area should be logged came up, especially after a 2006 article written by Oregon State

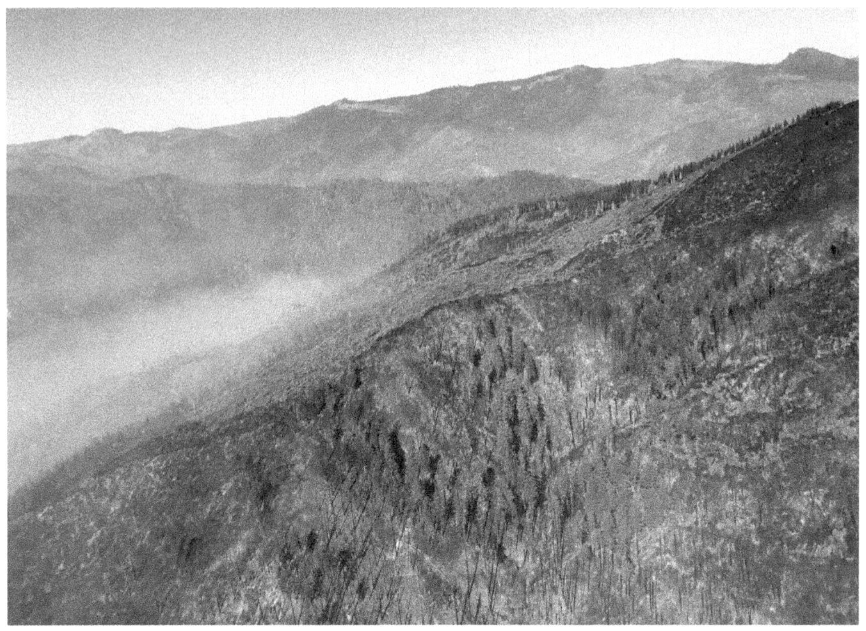

The area where the Chetco Bar Fire burned over the Biscuit Fire. *Courtesy of the U.S. Forest Service.*

University graduate student Dan Donato and several colleagues. Donato's evidence contradicted the USFS's views that this salvage logging helped the reforestation of the area. The Bush administration weighed in on the side of the USFS, and this put the environmentalists in a political and sometimes physical war with the federal and state government. This political fissure was never really settled, and in 2017, when the Chetco Bar Fire became the largest in Oregon, it burned in the same area, restarting this argument with some people.

The Toolbox Complex, 2002

Eleven days after the Biscuit Fire started, the Toolbox Complex was burning. Lightning started this fire as well, and just like the Biscuit Fire, the Toolbox Complex was a series of fires that burned together to form a fire that would go on to burn over 120,000 acres. While fighting this fire, one hotshot crew, one Type 2 AD (administratively decided) crew and a third Type 2 contract crew attempted a controlled burn operation. An AD crew is made up of people with needed skills and are usually veteran firefighters who are handpicked by fire administrative management.

Within an hour and a half, the controlled burn got out of control, and all the crews had to use their escape routes to get to the upper safety zone. The hotshot superintendent reassessed the situation, and after finding a dozer line to start another burn operation, he told the crews to proceed to a lower safety zone. The winds blew harder, and the fire grew in intensity. The hotshot superintendent instructed the hotshots and the AD crew to proceed to the burn, and since the contractor crew wasn't going to be a part of it, he told them to retreat to the upper safety zone. The contractor crew boss decided to scout the escape route to the upper safety zone by himself. In the time it took him to do that, the winds picked up even more, the smoke choked the contract crew and the extreme heat and wild fire behavior terrified them. By the time the crew boss returned, they feared for their lives. To calm their fears, the crew boss ordered them to deploy their shelters in the safety zone. This gave them a break from the heat and smoke for about fifteen minutes. Afterward, they were picked up, and eleven of the crew were medically treated and released. Two of them had minor burns, and the other nine had issues with smoke inhalation.[33]

No one ever wants to use a fire shelter. The people I've worked with dread the idea, but when nature goes wild and the fires rage, it can be completely terrifying, and it's better to be safe than dead. All around, 2002 was a scary year for wildland firefighters and expensive for the agencies fighting the fires. The Oregon Department of Forestry spent around $60 million, and the BLM and USFS spent $350 million. That year, 992,475 acres burned in Oregon, while 6 million acres burned across the country.

The South End Complex, 2006

This fire was also started by lightning, and from that one spark, it grew to burn over 117,000 acres. This was the 2006 season's biggest fire. It burned in the Steens Mountain and put the residents of Fields and Frenchglen on evacuation notice. The complex consisted of three smaller fires: the Pueblo Fire (approximately 78,000 acres), the Granddad Fire (35,000 acres) and the Krumbo Ridge Fire (around 3,300 acres).

THE EGLEY COMPLEX, 2007

The Egley Complex was the first conflagration of the year. It burned about ten miles from Riley, Oregon, but it only took two days to call off the conflagration, since there was no real threat to structures or life. All three of the main fires that made up this complex were started by lightning strikes. They eventually grew together to a size of 140,360 acres.

Through good fuels management and amazing work by the PNW Team 3 Type 1 Management Team on the west zone of the fire and the ORCA Type 2 Incident team on the east zone of the fire, the fire was controlled and finally put out without injury or loss of property.

THE HIGH CASCADES COMPLEX, 2011

Another thunderstorm sent a volley of lightning strikes into the Cascades, starting seven large fires that burned together, creating the High Cascades Complex, which burned over 108,000 acres. The fires grew because of wind gusts up to forty miles per hour and steady winds of twenty miles per hour. Again, the fuels across the Warm Springs reservation were very dry and easy to ignite. A rare thing occurred on this fire. A crew of fourteen from Linn County were caught inside a firenado while drafting water from Miller Creek into their structural fire engine. Many of the wildland firefighters, including Levi Lindsey from Albany and Kevin Rogers and Alex Parker from Brownsville, had to jump in the creek and lie face down in the water to escape the flames. Rogers told the *Oregonian*, "When the fire blew up, it was fire coming from three different sides."[34] Luckily, no one was injured.

The largest fire in the complex, the Razorback Fire, burned more than 51,000 acres and jumped the Deschutes River. The Powerline Fire and the West Hills Fire burned almost 16,000 acres each, while the Seekseequa Fire spread to about 7,000 acres around Lake Billy Chinook. There were several other small fires. This fire ended up costing $28,294,465. In 2011, all the Oregon fires burned over 315,000 acres, and the total cost was $91,529,866.[35]

The Long Draw Fire, 2012

The Long Draw Fire started on July 8, 2012, and was reported that evening at 1804 hours just southwest of Burns Junction. Within two hours, it went from a small spark to a raging sea of fire that spanned thousands of acres. In less than twenty-four hours, the fire covered 25,500 acres. On the second day, the fire burned 325,000 more acres, and on July 16, it consumed 558,052 acres of land, much of it BLM land that had been leased to local farmers for grazing.[36] The wind-driven fire swept through the high desert until it had consumed 871 square miles.

According to a BLM Fire Review of this incident, many people who were interviewed said, "I have never seen fires burn like this before." This is true because there hadn't been a fire of this magnitude in Oregon for over 150 years. To fight this fire, local crews worked with the Oregon Department of Forestry, BLM crews, the Rangeland Fire Protection Association (RFPA), the Oregon Department of Transportation and many other organizations. Remember, the RFPA is one of the nonprofit volunteer organizations supported by the Oregon Department of Forestry. The Jordan Valley RFPA fought this fire.

As one can imagine, fighting fire over such a large area can cause communication problems, especially with such a diverse group of organizations fighting the fire. Many people had to rely on their own cellphones as opposed

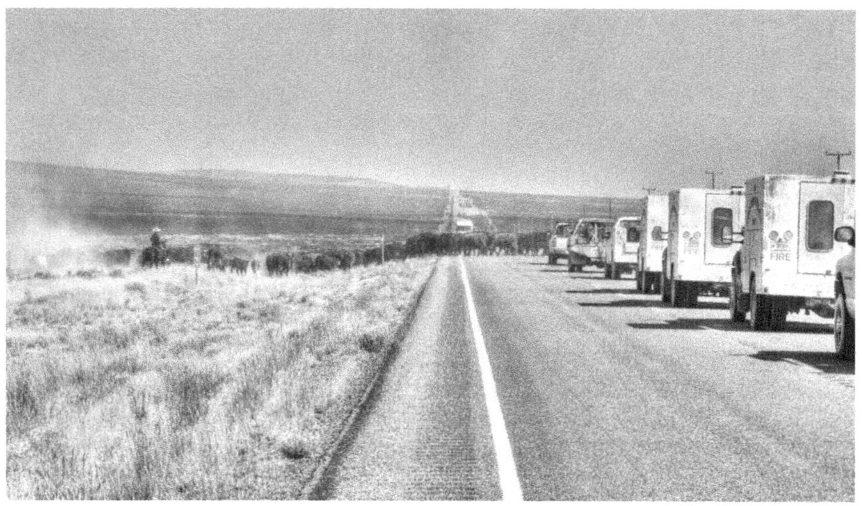

A local rancher drives his cattle to safety away from the Long Draw Fire. *Photograph by Kevin Abel, Oregon BLM.*

to radios. Another problem wildland firefighters had on this fire was the cattle. Imagine fighting a wildfire with thousands of head of cattle running from the fire. The fuels that were burning were also food for the cattle, in which many of the RFPA volunteers had stock. While the BLM and other crews tried to think strategically and wanted to use controlled burns, the RFPA personnel saw any additional burning as a detriment to their livelihood. The RFPA personnel were quoted as saying, "The time to really make headway is early in the morning. Teams and crews were still having meetings at 8:00 a.m. and not getting in place until it was too late." Another RFPA wildland firefighter said, "BLM fights fire, we put them out."

In the end, all the agencies worked together to put out this fire. BLM crews, RFPA crews, helicopters, engine crews, bulldozers and other heavy equipment, over a dozen water tenders and more all worked together to put out the largest fire Oregon had seen in over 150 years. In an interview with Oregon Public Broadcasting, resident Malena Konek said, "If it weren't for the firefighters holding their ground… my house wouldn't look much different from the charred brush just a hundred feet away."[37]

The Miller Homestead Fire, 2012

At the same time the Long Draw Fire burned, and only a little over one hundred miles away, the Miller Homestead Fire blazed to over 161,000 acres; 100,000 of those acres burned in one night. Over seventy cattle suffocated or burned to death, but the biggest losses, according to Stacy Davies, manager of Rolling Springs Ranch in Frenchglen, were the grazing lands they depended on to feed their cattle. The land affected with fire took three years to get back to where it was. In that time, cattle couldn't feed on that land.

Another great loss was the habitat for the sage-grouse. Eighty-five percent of the land burnt was sage-grouse habitat, and by some estimates, there were only 200,000 of the birds left in the United States. The Long Draw Complex and the Miller Homestead Fire resulted in the consideration of putting the sage-grouse on the endangered species list. Luckily, the U.S. Fish and Wildlife Service announced in a 2015 status review that the sage-grouse remains "relatively abundant and well-distributed across the species' 173 million acre range and does not face the risk of extinction."[38]

The BLM replanted almost 10,000 acres of sagebrush in the Miller Homestead Fire area and reseeded 25,500 acres. The price of grass seed coincidentally skyrocketed that year from two dollars per pound to eighteen dollars per pound. Sagebrush seed cost twenty-eight dollars per pound, and it takes up to four pounds to reseed an acre.[39] Multiply that by 25,500 acres; the cost of wildland fires isn't only measured in paying for the workers and equipment it takes to fight the fire.

The Holloway Fire, 2012

Seventy-three miles south of the Miller Homestead Fire, a fire started in Nevada and burned across the border into Oregon. It burned over 245,000 acres on the Oregon side of the border and an additional 215,000 acres in Nevada. All three of the fires involved were started by lightning strikes and burned through grass and scrubland.

All in all, 2012 marked the warmest year on record for the United States since 1895 according to the National Oceanic and Atmospheric Administration (NOAA). The acres burned per fire averaged 137.1, which was the most in recorded history. Before that, the average was 85.2 acres burned per fire. In Oregon, specifically, the Northwest Interagency Coordination Center's annual fire report for 2012 noted that Oregon had fewer fires than usual, but the acreage burned more than tripled the ten-year average, and that was only because 2006 and 2007 were such intense years. All of the fires in 2011 burned maybe 250,000 acres, half of the acreage burned in the Long Draw Fire. In 2012, Oregon had 2,935 fires and a total of 1,518,979 acres burned to ash.

The Buzzard Complex, 2014

The Buzzard Complex started about forty-five miles southeast of Burns and consumed 395,747 acres in the area of Stinking Water Creek Basin. This blaze was again caused by lightning, and like the fires just two years earlier, it burned through grass and sagebrush. The flames scattered a herd of 750

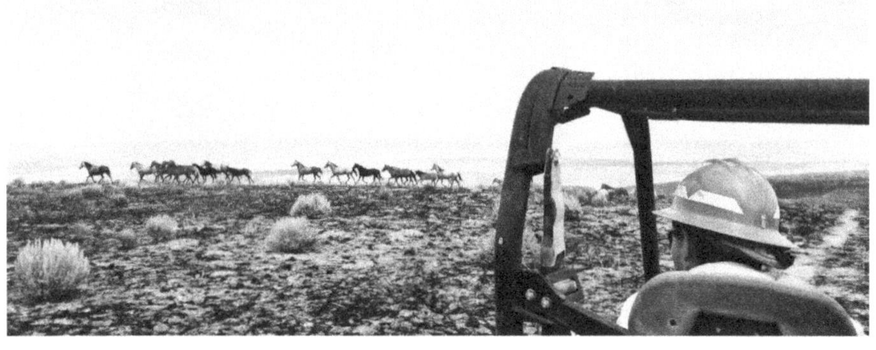

A wildland firefighter watches a herd of horses run across the blackened plain after the Buzzard Complex Fire, 2014. *Courtesy of the U.S. Forest Service.*

cattle and calves on the Drewsey Field Ranch. The Long Draw Complex still loomed in the memories of many Eastern Oregon residents. The *Oregonian* reported that one-hundred-year-old juniper trees "went up like Roman candles," leaving only a carpet of circular ash at the base. Combatting this fire required 234 wildland firefighters, 2 helicopters and 9 engines, but they did get to 100 percent containment.

THE CORNETT–WINDY RIDGE FIRE, 2015

The Windy Ridge Fire that caused residents from Alder Creek, Burnt River Canyon, Dry Creek and other areas to be evacuated merged with the Cornett Fire. The threat of the Cornett Fire caused the evacuation of Beaver Creek, Black Mountain, Denny Creek, French Gulch and more areas. There are three levels to wildland fire evacuation: level one lets people know that there is a possibility that, for their safety, they will have to pack what they can and leave their homes; level two is when they actually have to pack up their most valued possessions; and level three is when they have to leave their homes. Can you imagine having to leave your home and just hope that it will be there when you are allowed to come back after the danger has passed?

This complex burned timber and underbrush across almost 104,000 acres and threatened 187 single residences and 275 minor structures. Four homes and around a dozen minor structures were destroyed in this fire.[40]

THE CANYON CREEK COMPLEX, 2015

Lightning caused both of the largest fires in 2015. The Canyon Creek Complex started on August 12, 2015, one mile south of the towns of John Day and Canyon Creek. This fire didn't take a life, but it did destroy more private property than any other Oregon wildfire since the 1936 Bandon Fire. This fire raged through forty-three homes and destroyed nearly one hundred workshops, barns and satellite structures. The fire killed livestock and pets, destroyed farm equipment and vehicles and turned acres of timber to ash.

The *Oregonian* did a story on the mismanagement of this fire. The story accused the USFS of shipping off "three of their four 20-person crews in the days before the fire, despite dire weather predictions and staffing guidelines that required at least two crews remain."[41] This resulted in the incident commanders on the fire being understaffed during the crucial first hours of the fire.

Other complaints included landing smokejumpers and rappellers too far away from one of the fires that made up the complex, the incident commander sending crews to rest too early on the first night and deploying them too late on the second day, communications being "bungled" and evacuation notices going out too late for residents in Canyon Creek, "forcing many to flee with little notice and few possessions."

In an after-incident report, the USFS gave reasons for each complaint. First, as far as sending the crews away, the *Oregonian* may not have known about the interagency agreement. During a year as bad as 2015, resources do get spread pretty thin. The USFS is not only responsible for the entire state but the country as well. The report stated:

> *In Oregon and Washington, 15 Large Fires (a wildfire of 100 acres or more in timber or 300 acres or more in grass/sage) are active with 5,947 firefighters assigned. Across the nation, 62 active Large Fires are burning with 20,747 firefighting resources assigned. The severe lightning storm that impacted Malheur National Forest area on August 12, also establishes new fire starts throughout the region—creating an additional draw of firefighting resources. No forest is able to staff for the worst case scenario. Our staffing levels are based on our budget and augmented with severity funding from the regional office when fire danger is very high. The interagency wildland fire community shares resources across the country by sending available firefighting personnel and equipment from the slower areas to the region that is experiencing the most severe fire conditions.*[42]

The USFS also said that the reason why the smokejumpers and rappellers were late getting to the fire is because morning thunderstorms had rolled in, making conditions hazardous for flying into the area. The pilots in both fixed-wing planes and helicopters finally opted to fly anyway, but safety conditions had grounded them for an hour and a half.

There are always lessons to be learned after a fire, and no agency, organization or crew will fight a fire without doing something they can improve on afterward, but everyone on a fire—no exceptions—believes that the safety of the firefighters comes first. As far as the USFS goes, half of its entire budget goes to fighting fires, although that is only maybe 25 percent of the USFS's original mission, and in 2018, the current administration is cutting the budget even more. We should ask ourselves if reacting to raging fires is a better strategy than preventing them from happening in the first place. However, one problem with this strategy is that if the USFS, BLM, Oregon Department of Forestry and other agencies are successful in preventing wildland fires, people will ask why they have to spend so much tax money on preventing fires when there are no incidents.

EAGLE CREEK FIRE, 2017

I want to express how sorry I am for what I did. I know a lot of people suffered because of a bad decision that I made. I'm sorry to the first responders who risked their lives to put out the fires, I am sorry to the hikers that were trapped, I am sorry to the people who worried about their safety and their homes that day, and for weeks afterwards. I am truly sorry about the loss of nature that occurred because of my careless action. Every day I think about this terrible decision and its awful consequences. Every time I hear people talk about the fire, I put myself down. I know I will have to live with my bad decision for the rest of my life, but I have learned from this experience and will work hard to help rebuild the community in any way that I can. I now realize how important it is to think before acting because my actions can have serious consequences. I, myself, love spending time in nature and now I realize how much work it takes to maintain the National Forest so people can enjoy it. I sincerely apologize to everyone who had to deal with this fire, I cannot imagine how scary it must have been for you. I know I have to earn your forgiveness and I will work hard to do so and one

> *day, I hope I will. Thank you for giving me a chance to speak. This has been a big learning experience for me and I take it very seriously. I apologize with all my heart to everyone in the gorge.*[43]

These are the words of the fifteen-year-old boy who threw two firecrackers into Eagle Creek Canyon on September 2, 2017. The first firecracker exploded in the air. The second firecracker led to a blaze that torched 3,000 acres overnight, and in only three days, it burned 20,000 acres and spotted across the Columbia River into Washington. The fire doubled in size, and by the end, it reached nearly 50,000 acres; changed the landscape of the Columbia River Gorge; reduced the air quality for Portland, the largest city in Oregon; forced hundreds of people to evacuate; and shut down a major artery (Interstate 84) for days. A firecracker that could have been bought with the change you find in your couch ended up costing taxpayers upwards of $30 million—and that isn't counting the commercial money loss from shutting down semitrailers on Interstate 84.[44]

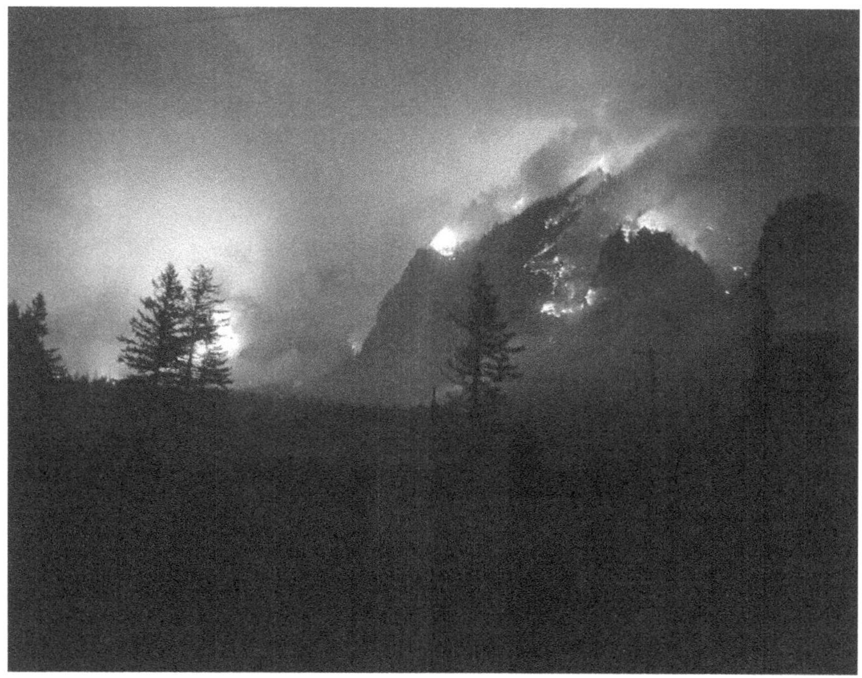

The Eagle Creek Fire burning at night, 2017. *Courtesy of the U.S. Forest Service.*

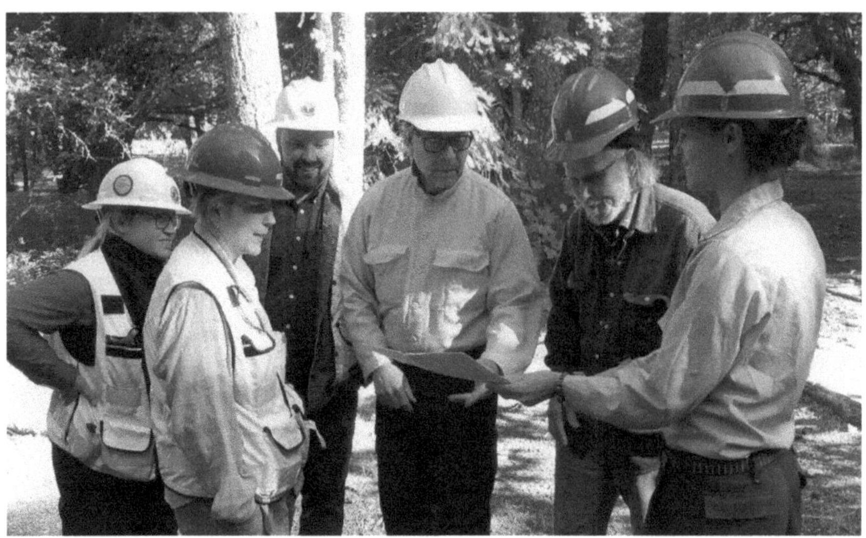

This is the Eagle Creek BAER (Burned Area Emergency Response) Team. They were assigned to this fire to create a Soil Burn Severity map. The BAER Team will "identify the areas likely to be impacted after the fire. Potential treatments to mitigate threats to 'Values at Risk', including human life, structures, roads, trails, streams, cultural resources, and other natural resources, will be evaluated." *Courtesy of the U.S. Forest Service.*

This fire didn't burn as many acres as the other fires in this chapter, but the Eagle Creek Fire was all over the news because of its location and how it started. On a fire in an area such as the Columbia River Gorge National Scenic Area, putting out the fire is only phase one. The cost noted above also doesn't count sending in archaeologists, erosion specialists and other experts who can gauge whether the area is safe for future hikers. Many of the trails have to be rebuilt and restored, and these efforts can take years. Private companies such as Full Sail Brewing Company, the Oregon Community Foundation, Friends of the Columbia Gorge, Double Mountain Brewery and more put together a crowdfunding effort called "Be There for The Gorge" that raised over $148,000 to help restore the trails in the Gorge that were damaged by this fire. If you are interested in contributing, you can visit www.nationalforests.org and click the "donate" button.

Chetco Bar Fire, 2017

The Chetco Bar Fire started with a lightning strike in the Kalmiopsis Wilderness on July 12. The lightning struck between Cave Junction and Brookings. The fire spread to over 191,000 acres. Many people criticized the way the USFS handled this fire. Some went so far as to say that the Chetco Bar fire was allowed to burn as a part of some sort of "liberal agenda."

This fire started in a very remote part of the Kalmiopsis Wilderness. The first wildland firefighters to respond were members of a helitack team that had to rappel in. They struggled with the steep and unstable slopes. In that sort of situation, there is no escape route that a firefighter can take and hope to outrun the fire. Not only that, but no one could find a good place for a safety zone. The incident commander made the call to not put the wildland firefighters in that type of danger.

Civilians complained that if the USFS would have just dumped a few buckets of water on the small fire, it would have gone out, but as has been

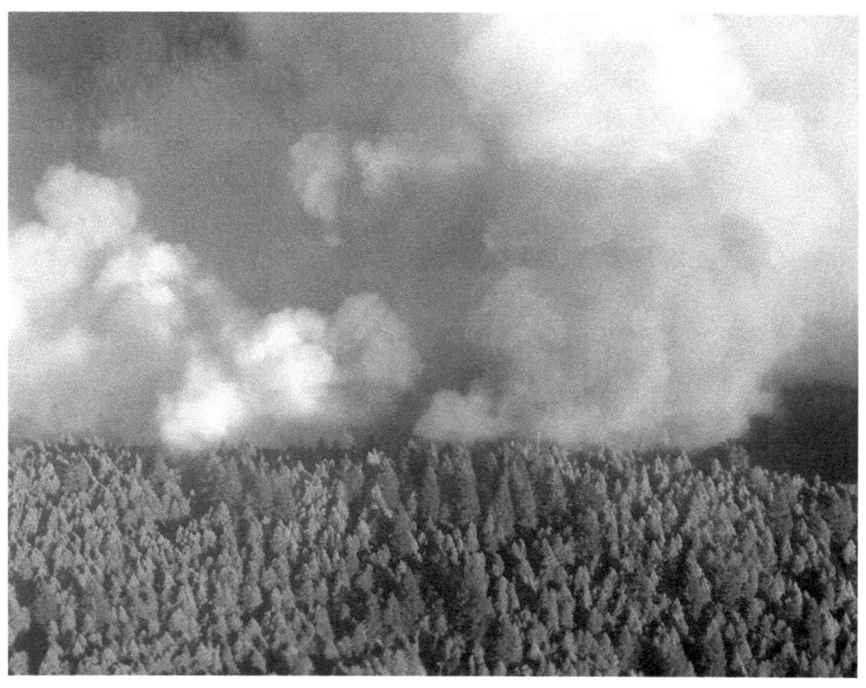

The Chetco Bar Fire burning from the air, 2017. *Courtesy of the U.S. Forest Service.*

stated multiple times in this book, the only way to put out a fire is with boots on the ground and tools in hand. Incident commander Monty Edwards met with the public and told them, "They can't do it alone with just aviation assets. You need someone [on] the ground to go in direct and knock out the fire on the ground. That's basic firefighting."[45] In fact, during the first day the fire burned, helicopters dropped seventeen thousand gallons of water on this fire—that's fifty buckets with the helicopter they had available.

The plan was to pull back and cut a fire line between the fire and the town of Brookings, but the wind shifted. Not only did the winds shift, they picked up to speeds that were completely unexpected. On August 15, the fire blew up. The forty-five-miles-per-hour winds funneled it down the Chetco River Valley. The blaze easily jumped the Chetco River, and the containment line meant to protect Brookings and the fire grew from 6,000 acres to 91,000 acres. It was making a run that caused the call for evacuations. The fire dimmed the sky during the day, choked people when they walked outside and burned through timber. Residents throughout Southern Oregon were scared, angry and fed up, and they demanded that their questions be answered. Curry County commissioner Court Boice said that something needed to be changed. He said, "I don't know how much more wildfire we can handle. We're losing our timber and our watersheds."

An article in the *Seattle Times* written by Jeff Duewel, a journalist with the *Grants Pass Daily Courier*, reported that a person in a town hall–type informational gathering asked if the USFS had learned anything since the Biscuit Fire. Deputy forest supervisor Craig Trulock replied, "What we've learned has been offset by climate change and lots of fuel buildup, and we have more severe fire than ever before."[46]

Chapter 8

WHY THEY FIGHT

When you ask a wildland firefighter why he or she fights fire, he or she may not have an answer. They only know that they love this dangerous and difficult job. Many first-year wildland firefighters will quit sometime before their first season is over, but the majority of the ones who stick it out will make a career out of it. Once you're hooked, you're hooked. It's not something you can easily explain, but that same spirit has lived in wildland firefighters for decades. The love for the job doesn't stop with the wildland firefighter. Many times, they will bring in their friends and family members.

Here are a few guesses as to why people love this dangerous job. First, wildland firefighting is a rewarding job. The physicality of the job can help a person's overall mental well-being. The Mayo Clinic wrote a study stating, "Exercise helps prevent and improve a number of health problems, including high blood pressure, diabetes and arthritis. Research on depression, anxiety and exercise shows that the psychological and physical benefits of exercise can also help improve mood and reduce anxiety."[47] While it seems like working in a dangerous field where you could be injured or killed may increase anxiety and hurt your mental well-being, the opposite is often true.

What's more, being in nature is proven to help well-being. A Stanford-led study found quantifiable evidence that nature helps fight anxiety and depression. The coauthor of the study and senior fellow at the Stanford Woods Institute for the Environment, Gretchen Daily, said, "These results suggest that accessible natural areas may be vital for mental health in our rapidly urbanizing world." The study found that people who live in a urban environment have a 20 percent higher risk of anxiety disorders and a 40 percent higher risk of mood disorders compared to people in rural areas.[48]

Wildland firefighters not only have the benefits of exercise and nature but are also working hard to preserve our wildlands, prevent property damage and save lives. They have a worthwhile purpose, which may be lacking in many of the careers people choose today.

Wildland firefighting has all these benefits plus the camaraderie, excitement and adrenaline rush of danger and the opportunity to lead others. In this way, wildland firefighting is very similar to the military. In fact, if you were to read the IRPG (Incident Response Pocket Guide), a small spiral-bound pocket book with all the information you need on a wildland fire, you would see that the first several pages are about duty, respect and integrity. This is all taken from the eleven principles of leadership learned in the military. The operations order from the IAP (Incident Action Plan), the daily packet handed out at leadership meetings along with the general strategy to control the fire, is based on the army's operation order. The risk assessments, incident size-up, chain of command, "9 line" medical

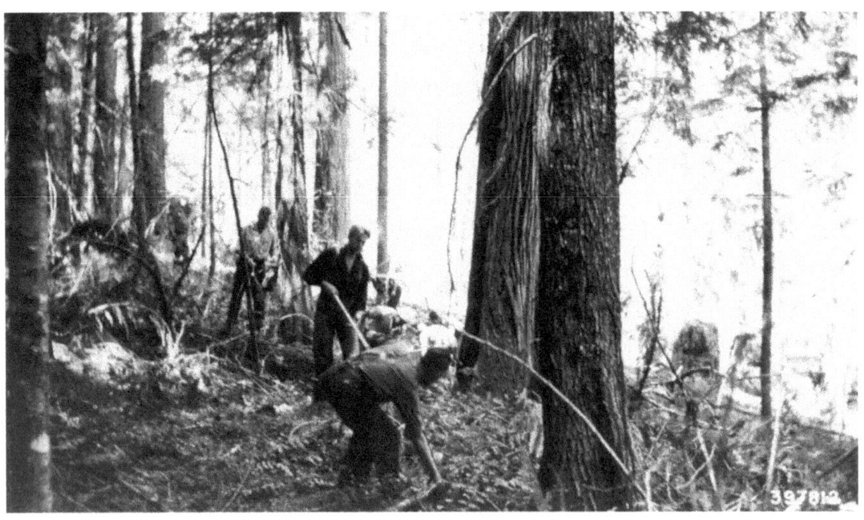

The Willamette Flying Twenty training in June 1940 on the progressive method of fire line construction still in use today. *Photograph by Roy A. Elliott, courtesy of the Oregon Department of Forestry.*

evacuation and many other standard operating procedures are based on the policies and procedures of the military.

Many veterans feel right at home fighting wildland fires. I will interject myself into this part of the book only because I don't want to speak for all the veterans out there, but as a combat veteran with a Purple Heart, I think being on a large fire is very similar to being deployed. It has all the best parts of being involved in a real-world incident without getting shot at or blown up. I've seen just as much selflessness and courage on the fire line as I saw at war, and the people we are risking our lives to protect live here in our communities and are very appreciative for what we do.

All of this aside, the biggest reason I've seen other wildland firefighters do this amazing and difficult job is the reward of helping others. Maybe you were born in Oregon, you've lived here for years or you just moved here and found you've never belonged anywhere else—the wilderness, the forests and wildlife are in your DNA. For those of us who have a hard time working indoors and want an "office" with a view of trees in every direction, fresh air and the satisfaction of a hard day's work, being a wildland firefighter is the perfect profession, and it's a profession we will need more and more as the years pass.

Wildland firefighting is more than a job. Visit a fire camp and you will find the best parts of America. Men and women of all races, religions and backgrounds work together to help save the lives and property of complete strangers, and no matter where a wildland firefighter is from or what their role is on the fire, we're all brothers and sisters. We're all a part of a something larger than ourselves. Sometimes we work months straight, sixteen-hour days, with only a few hours of downtime scattered throughout the summer, and it is some of the hardest work a person can do. But most, like myself, if asked, wouldn't trade it for the world. There's a reason we keep coming back and bringing our friends and family members to be a part of it. My son will be fighting fire with me for his third year this coming summer. There is danger, but the benefits outweigh any of it. It's an honor to protect the land and work to save lives and property in our wildlands.

NOTES

Introduction

1. William Clark, *The Journals of the Lewis & Clark Expedition* (Lincoln: University of Nebraska Press, 2001), Lewisandclarkjournals.unl.edu.
2. Phil Cheney and Andrew Sullivan, *Grassfires: Fuel, Weather and Fire Behaviour*, 2nd ed (Australia: CSIRO Publishing, 2008).

Chapter 1

3. Joy Drohan, "Exploring Patterns of Burn Severity in the Biscuit Fire in Southwestern Oregon," *Fire Science Brief* (January 2010, Issue 88), http://digitalcommons.unl.edu/jfspbriefs/71/.
4. "Forest Facts," Oregon Department of Forestry (Salem: State of Oregon, 2009).
5. "Wildfire Smoke and Your Health," U.S. Forest Service, https://www.fs.usda.gov/Internet/FSE_DOCUMENTS/stelprdb5318238.pdf.

Chapter 2

6. "About the Agency," U.S. Forest Service, 2017, https://www.fs.fed.us/about-agency.
7. James Gersbach, personal interview, 2017.
8. Ariel Scotti, "President Trump's Proposed Budget Cuts to Fighting Wildfires Could Be Deadly for the Western U.S.," *New York Daily News*, July 11, 2017, https://www.nydailynews.com/news/national/trump-2018-budget-cuts-fighting-wildfires-deadly-article-1.3318296.

Chapter 3

9. Sam Swetland, personal interview, 2017.

Chapter 4

10. Leslie Anderson, "The New Generation Fire Shelter," U.S. Department of the Interior, National Wildfire Coordination Group, 2003, https://www.nwcg.gov/sites/default/files/publications/pms411.pdf.
11. Historical Wildland Firefighter Fatality Report by Year, 1910–2016, National Interagency Fire Center, https://www.nifc.gov/safety/safety_documents/Fatalities-by-Year.pdf.
12. "Smokejumpers," U.S. Forest Service, 2017, https://www.fs.fed.us/fire/people/smokejumpers.
13. "Redmond Smokejumpers," U.S. Forest Service, 2015, https://www.fs.fed.us/fire/people/smokejumpers/RAC/history.html.

Chapter 5

14. Warren Cornwall, "Who Is Starting All Those Wildfires? We Are," *Science*, September 12, 2017, https://www.sciencemag.org/news/2017/09/who-starting-all-those-wildfires-we-are.
15. "Wildfire Causes and Evaluations," National Park Service, 2017, https://www.nps.gov/articles/wildfire-causes-and-evaluations.htm.

Chapter 6

16. "Controlled Burning," U.S. Forest Service, 2017, https://www.fs.usda.gov/detail/dbnf/home/?cid=stelprdb5281464.
17. "Coastal Vegetation and Plant Ecology," Oregon Explorer, Corvallis, OR: OSU Libraries, 2017, http://oregonexplorer.info/content/coastal-vegetation-and-plant-ecology.
18. "Find a Cabin," U.S. Forest Service, https://www.fs.usda.gov/detail/r6/recreation/?cid=stelprdb5290342.
19. "Forest Resources," Oregon Climate Change Research Institute, 2017, http://www.occri.net/pnw-impacts/forest-resources/.
20. Meghan Dalton, Kathie Dello, et al., *The Third Oregon Climate Assessment Report*, Oregon Climate Change Research Institute (Corvallis, OR: Oregon State University, 2017).

21. Fiscal Year 2017 Budget Overview, U.S. Department of Agriculture, U.S. Forest Service, 2016, https://www.fs.fed.us/sites/default/files/fy-2017-fs-budget-overview.pdf.

Chapter 7

22. Leonard Whitmore, "The Great Forest Fire," 1986, http://cliffhanger76.tripod.com/c2sea/fire/index.html.
23. "Pacific City, Oregon," Revolvy.com, https://www.revolvy.com/main/index.php?s=Pacific%20City,%20Oregon&item_type=topic.
24. David Wilma, "1902 Yacolt Burn, Washington," 2011, http://columbiariverimages.com/Regions/Places/yacolt_burn_1902.html.
25. Timothy Egan, *The Big Burn: Teddy Roosevelt and the Fire That Saved America* (New York: Mariner Books, 2010).
26. "Tillamook Burn," https://oregonencyclopedia.org/articles/tillamook_burn/#.Wph71ZOplBx.
27. Oregon History, Twitter post, September 6, 2017, 8:12 p.m., https://twitter.com/oregon_history.
28. "Invasion and Inferno: The Story of the Bandon Fire," 2010, http://oregoninvasivespecies.blogspot.com/2010/04/invasion-and-inferno-story-of-bandon.html.
29. James Sinks, "Residents Pay For Fires on Unprotected Properties," *Bend Bulletin*, August 24, 2005.
30. "Incident Management Situation Report," National Interagency Fire Center, 1996, https://www.predictiveservices.nifc.gov/IMSR/1996/19960817IMSR.pdf.
31. The Emergency Conflagration Act, Oregon Revised Statute 476.510 to ORS 476.610.
32. "The Biscuit Fire: Time to Bury the Myths," 2012, https://kalmiopsiswild.org/1019/the-biscuit-fire-time-to-bury-the-myths.
33. "Toolbox Deployment Incident Overview," 2002, https://www.wildfirelessons.net/HigherLogic/System/DownloadDocumentFile.ashx?DocumentFileKey=a51fa2d2-341a-44eb-86e9-e8a3cf44e628.
34. Kimberly A.C. Wilson, "High Cascades Complex Fires Continue to Burn More Than 90,000 Acres in Central Oregon," *Oregonian*/OregonLive, September 2, 2011.
35. Northwest Annual Fire Report, Northwest Interagency Coordination Center, 2011, https://gacc.nifc.gov/nwcc/content/pdfs/archives/2011_Annual_Report.pdf.

36. Long Draw/Miller Homestead Fire Review, Bureau of Land Management, 2013, https://www.blm.gov/or/news/files/long-draw.pdf.
37. David Nogueras, "Long Draw Fire Leaves Wake of Damage," Oregon Public Broadcasting, July 13, 2012, https://www.opb.org/news/article/long-draw-fire-leaves-wake-damage/.
38. "Greater Sage-Grouse," Oregon Department of Fish and Wildlife, http://www.dfw.state.or.us/wildlife/sagegrouse.
39. Richard Cockle, "Range Burned by Long Draw, Other Huge Wildfires May Recover More Quickly Than Expected," *Oregonian*/OregonLive, October 22, 2012.
40. "Cornet–Windy Ridge Fire News Release," National Wildfire Coordination Group, September 26, 2015, https://inciweb.nwcg.gov/incident/article/4478/30151.
41. Laura Gunderson and Ted Sickinger, "Burned: Firestorm," *Oregonian*/OregonLive, August 12, 2016, http://www.oregonlive.com/wildfires/index.ssf/page/canyon_creek_fire.html.
42. "Canyon Creek Complex," U.S. Forest Service, 2016. https://www.fs.usda.gov/Internet/FSE_DOCUMENTS/fseprd503421.pdf.
43. Maxine Bernstein, "Teen Admits Starting Eagle Creek Fire, Sentenced to 5 Years of Probation," *Oregonian*/OregonLive, February 16, 2018.
44. Ericka Cruz Guevarra, "How Much Has the Eagle Creek Fire Cost and Who's Paying?," Oregon Public Broadcasting, October 20, 2017, https://www.opb.org/news/series/wildfires/oregon-eagle-creek-fire-price-tag-payment.
45. Liam Moriarty, "Could the Chetco Bar Fire Have Been Prevented?," Oregon Public Broadcasting, October 9, 2017, https://www.opb.org/news/article/chetco-bar-wildfire-preventable.
46. Jeff Duewel, "Forest Officials in Oregon Explain Fire Strategy to Skeptical Public," *Seattle Times*, October 2, 2017, https://www.seattletimes.com/seattle-news/forest-officials-in-oregon-explain-fire-strategy-to-skeptical-public/.

Chapter 8

47. "Depression and Anxiety: Exercise Eases Symptoms," The Mayo Clinic, 2017.
48. Rob Jordan, "Stanford Researchers Find Mental Health Prescription: Nature," *Stanford News*, June 30, 2015, https://news.stanford.edu/2015/06/30/hiking-mental-health-063015.

ABOUT THE AUTHOR

Sean Davis is the author of *The Wax Bullet War*, a book that chronicles his experiences in the Iraq War and the Hurricane Katrina cleanup. He was awarded the Purple Heart for injuries sustained while in combat. Davis won the Legionnaire of the Year Award from the American Legion in 2015, and he was also the recipient of the Emily G. Gottfried Human Rights Emerging Leader Award in 2016. He was also chosen by KATU as the newsmaker of the year in 2016 and was knighted in Portland by the Royal Rosarians. His stories, essays and articles have appeared in Lidia Yuknavitch's book *The Misfit's Manifesto* (Simon & Schuster), the Forest Avenue Press bestselling anthology *City of Weird*, *HUMAN* (a film by Yann Arthus-Bertrand) and on *60 Minutes*, The Big Smoke and many more. Davis teaches writing and literature classes in colleges around Portland and at the William Joiner Institute for the Study of War and Social Consequences at the University of Massachusetts, Boston. In the summer, he heads out on a wildland engine to fight fires on the West Coast. He currently lives in McKenzie Bridge, Oregon, with his beautiful wife, Kelly Davis; his incredible daughter; his dog, Bullet; and the family's chickens.

www.ingramcontent.com/pod-product-compliance
Lightning Source LLC
LaVergne TN
LVHW052338100826
845147LV00020B/1100

* 9 7 8 1 4 6 7 1 3 8 5 0 5 *